The Way of Coyote

THE WAY OF
coyote

Shared Journeys in the Urban Wilds

GAVIN VAN HORN

The University of Chicago Press

Chicago and London

The University of Chicago Press, Chicago 60637
The University of Chicago Press, Ltd., London
© 2018 by Gavin Van Horn
Published 2018
Paperback edition 2025

34 33 32 31 30 29 28 27 26 25   1 2 3 4 5

ISBN-13: 978-0-226-44158-0 (cloth)
ISBN-13: 978-0-226-84011-6 (paper)
ISBN-13: 978-0-226-44161-0 (e-book)
DOI: https://doi.org/10.7208/chicago/9780226441610.001.0001

Illustrations by Keara McGraw

Library of Congress Cataloging-in-Publication Data

Names: Van Horn, Gavin, author.
Title: The way of coyote: shared journeys in the urban wilds / Gavin Van Horn.
Description: Chicago: The University of Chicago Press, 2018. | Includes
    bibliographical references.
Identifiers: LCCN 2018013996 | ISBN 9780226441580 (cloth) | ISBN
    9780226441610 (e-book)
Subjects: LCSH: Urban animals—Illinois—Chicago. | Coyote—Illinois—
    Chicago. | Human—animal relationships—Illinois—Chicago. | Animals
    and civilization—Illinois—Chicago.
Classification: LCC QH541.5.C6 V37 2018 | DDC 591.75/60977311—dc23
LC record available at https://lccn.loc.gov/2018013996

CONTENTS

A Companionable Dissolution to Plan A

> In a culture that hopes to stick around for a while, is it too much to ask its members to get to know a few of their nonhuman neighbors?
>
> —ROBERT MICHAEL PYLE, in *The Way of Natural History*[1]

Plan A failed to materialize as I envisioned. I ended up in a city. A big city. I wasn't supposed to end up in a city. Any city. I wasn't supposed to look out a condominium window and gaze upon a brick wall rather than a juniper woodland or silver-streamed valley. When I see maps advertising the reach of the latest cell phone coverage, I am most drawn to the uncolored blotches—the places where the call will be dropped, where even the cell phone providers shrugged their shoulders and determined, "Not worth it." I imagine those spots of cellular incognita as the places where the ratio of human to nonhuman is just about right.

Chicago splintered the logs of my fictive cabin in the woods. To its credit, Chicago built something else—a more unified perception:

I've discovered that stupendous varieties of life can flourish in human-dominated landscapes.

What once in my mind was a biotic blank slate is now crisscrossed by the tracks, burrows, flyways, hunting territories, and migratory routes of other creatures who call the city home. Black-crowned night herons, peregrine falcons, garter snakes, red admiral butterflies, beavers, eastern gray squirrels, Cooper's hawks, bumblebees, hawk moths, chimney swifts, meadow voles, spring peepers, coyotes, opossums, nightjars, striped skunks, herring gulls, raccoons, white-tailed deer, little brown bats, and so many other creatures with other-than-human intelligences are moving among us, dwelling in our midst, and threading their lives through our own.

This is not to say city living isn't challenging. The footprint of Chicago, like other cities, is wide and deep. In a two-hundred-year stretch of time, the lands and waters in the Chicago region have been dramatically altered—wetlands drained and filled, rivers straightened and their flows rejiggered, living soils inundated by massive amounts of concrete and asphalt, and forests and prairies chewed apart by the grinding teeth of human enterprise.

The speed with which these changes occurred remains difficult to wrap one's mind around. In 1830, there were about one hundred people settled near the mouth of the Chicago River; now there are about nine and a half million in the greater Chicago metropolitan area. In the two decades between 1890 and 1910 alone, the urban population ballooned from one million to two million people, leading the historian Carl Smith to declare Chicago "America's most remarkable urban phenomenon."[2]

What would a time-traveling seventeenth-century French fur trapper or a native Potawatomi person thrust into a future of skyscrapers and gridlocked automobiles make of this modern cityscape? In the half blink of a geological eye, the ecological fabric of the Chicago region experienced an almost total overhaul.[3] Species populations changed dramatically in the process, with the weight shifting toward one in particular, *Homo sapiens*.

Chicago is a cauldron of relative newcomers and their descendants, people immigrating in waves, fleeing disenfranchisement and loss of homeland: Irish refugees escaping the potato famine in the mid-nineteenth century, Central and Eastern Europeans looking for skilled labor positions and political asylum, African Americans hoping to rise from poverty and post–Civil War segregation in the American South, and, more recently, Mexicans and Southeast and East Asian populations brought by prospects of a better life. The city reached its peak population—3.6 million—by 1950, and like many American cities in the post–World War II era, Chicago saw its suburbs explode while its urban core slowly depopulated.

Chicago has entered a postindustrial phase, its former identity as a Rust Belt powerhouse diminished by global manufacturing markets and the ebb of labor unions' strength. Yet change and reinvention are part of this city's social fabric. Since the 1990s, a revitalized downtown and continued restoration of urban natural areas—such as the lakeshore, large swaths of protected forest preserves and parklands, and the river that flows through the heart of the city—provide new reasons for another wave of people to turn cityward. The city promises, as it often has, a place to begin again.[4]

Cities throughout history—and Chicago is no exception—can clearly be places of extremes: pollution, habitat destruction, poverty, inequity. If you associate the word *urban* with the absence of nature, however, then you can't fully understand Chicago. The stories in this book occasionally lament lost abundance, but more often they point toward potential—a future in which urban dwellers take their cues from landscapes past and present to reach toward a more vibrant collective of ecological citizens. I've come to believe that the hope of Chicago—or any city, for that matter—lies with living into that belonging.

. . .

Three presences pop up on more than one occasion throughout this book. They are here because each has helped me make sense of the city, orienting me in the urban landscape. What they share in common: a disdain for boundaries.

I suspect Aldo Leopold (1887–1948), a wildlife ecologist with a pen as sharp as his mind, will raise the fewest eyebrows among readers.

Leopold famously wrote that nothing as important as a land ethic—the caring response of humans as "plain members and citizens" in the land community—is ever written. If such an ethic exists or will exist, like the unnamable Tao, one will never get hold of it precisely. As he puts it, the land ethic "evolves in the minds of the thinking community." The evidence for a land ethic is inscribed on the landscape, made manifest in the ways we engage with the lands (and waters, and airs) of which we find ourselves a part. *The Way of Coyote* constitutes my effort to suss out an urban land ethic.

I follow many footpaths while considering the contours of an urban land ethic, some across cement, some through wetlands, some between the sturdy stalks of prairie grasses. These paths interweave with reflections on connection, disconnection, and how to mend broken lifelines. I believe when a person can freely walk, when there is a safe path to place the next footstep, it is a decent indicator that the landscape is alive. It's that simple. When for some reason we cannot, when there are barriers to one of the most basic acts of mobility, something has gone awry. And we need to ask why.

Leopold was a person who asked why. His intuitions and experiences in forestry and wildlife ecology taught him that social and ecological health were linked, and he sought ways to encourage that perception. He often counseled his readers, students, and colleagues to cease thinking in terms that pitted agriculture against wilderness, private landowner against federal government, and emotional attachment against rational management. His professional work and personal commitments became a series of integrations. Over the course of his life, he sought to better link the evolving landscape of his thought with what he saw occurring in the actual landscapes of the Midwest, and Wisconsin in particular, but also more broadly in Arizona, New Mexico, northern Mexico and Baja California, Germany and Canada, among other regions.[5]

In his writings, fewer insights about cities can be found than the major conservation concerns of his (and still our) time—soil loss, species eradication, and habitat fragmentation. And yet his thinking was too broad not to turn an occasional eye toward the city, as he did when he issued this stunning statement, particularly pertinent to the urban wilds: "The weeds in a city lot convey the same lesson as the redwoods; the farmer may see in his cow-pasture what may not be vouchsafed to the scientist adventuring in the South Seas."[6] No matter where we find ourselves, our perception can be attuned to such instances of wildness, and we need not venture from our backyards to fall in love outward.

Leopold believed in attention to nature's details. He kept extensive phenological records, filling handwritten journals with the timing of spring flower blooms and birdcalls at twilight. He was not a one-dimensional observer, standing back and apart from the land, teasing apart deductions in a laboratory with an interest only in so-called objective results. For Leopold, the land was alive and humans had a "vital relation" to it; in a word, the land was worthy of love.[7] City and country, this is a connection worth seeking—indeed, our attention and care unites landscapes and discloses the wild continuum we share with nonhuman others.

. . .

Lao Tzu, or, more accurately, the spirit of Lao Tzu, figures as my second companion on this urban journey. His presence may require a pinch more explanation. To be honest, I'm surprised he's here. Neither he nor the Taoist point of view that he represents is a typical dinner-table discussion topic (at least not at my family's dinner table), so I need to account for his presence in this book.

Lost in the mists of time is the person (or persons) who wrote the *Tao Te Ching*. Lao Tzu, if indeed he was a historical person, is the fellow most often given the credit. Lao Tzu is said to have lived in predynastic China, in the sixth century BCE, and served as the librarian in the court of Zhou. Lao Tzu was, thus, a city boy. According to legend, when the time came for him to retire, the pull of wilder country and his wish to live as a hermit drew him to the city's border. It was there that a gatekeeper recognized the sagacious man and begged him to transfer his thoughts into writing. More than 2,500 years later, the *Tao Te Ching* remains with us, even as Lao Tzu's further expeditions remain a mystery.

I'm not a Taoist practitioner in any formal sense, but I've found the *Tao Te Ching* a helpful interpretative tool for making sense of the city, for reconciling what seem to be dualities, such as the urban and the wild, into a more continuous whole. I think it's high time to welcome Lao Tzu back to town. So, I'm dragging him down from the misty mountains—at least in the form of his magnum opus, the *Tao Te Ching*.

As cloud shadows pass over an ocean, or prairie grasses bow to the breeze, the *Tao Te Ching* is a Rorschach test: the ink blots change with translation and with what a person brings to the text. A series of eighty-one poems, the *Tao Te Ching* offers a concision of words that opens boundless landscapes of imagination, creating breathing room for contemplative dialogue. Similar to the clouded ocean and the undulating prairie, I've found that the attention I give to this text is its own reward.

Thinking about the city through the lens of the Tao makes sense to me: analogous to the black-and-white circle of the yin and yang, with a dot of white in the black and a dot of black in the white, the city should neither be vilified nor championed. We need to understand the city in the country and the country in the city, for they create each other, and perhaps one of the great tasks of our time is to achieve a more harmonious unity between them. So sometimes you'll see a verse from the *Tao Te Ching* bend its way into an essay, offering a different way of thinking about the tenuous boundaries we often draw between humans and nature, urban and rural, and civility and wildness.

■ ■ ■

Once he knew the book's title would carry his name, the third character kept insisting on squeezing himself between the pages at key moments.

There is coyote. There is also Coyote. There is coyote, the flesh-and-blood animal, the urban adapter, the resourceful canid who has ably spread in numbers across North and Central America. One hundred years ago, in the United States, coyotes were associated almost exclusively with western deserts and the open prairies of the country's interior. Unlike other of their midsize and large-bodied mammal brethren, however, coyotes seem to grow stronger—or at least more prolific—the more they are persecuted. A good number of them have now determined they will take their chances in the city.

To me, coyotes are iconic urban adapters. I believe we all will be better off when we learn how to live with them. It may be that we have no choice but to do so. The decision is as much theirs as ours, for although they are always on the move and often out of sight, they are here to stay. The city is now coyote's.

Which brings us to Coyote. In myth and legend, Coyote is a venerated person, particularly among Native peoples of the American West, from the Salish in Washington State to the Navajo in Arizona. And still further afield. As Dan Flores notes in *Coyote America*, Coyote "is the most ancient deity of which we have record on this continent," and stories about Coyote (or "Old Man Coyote" as he is sometimes known) are likewise among the oldest in North America. "No other native deity in America came anywhere close to inspiring such a vast body of oral literature," Flores observes. "West of the Mississippi, across the last 10,000 years, Coyote has been America's universal deity, surviving as a Paleolithic god among agricultural peoples like the Wichitas and ultimately reaching as far south as the Aztecs, who knew him as Huehuecoyotl, Old Man Coyote. Or Old Man America."[8]

Coyote is a piece of work, *the* trickster figure—comical, bawdy, occasionally helpful, consistently entertaining. The lead essay in this book is about flesh-and-blood coyotes, but it is Coyote as trickster, the mischief maker, who felt compelled to nose his way into various sections of this book, parting the pages as he would little bluestem grasses in order to have a look around. This Coyote is my attempt to give a time-honored myth new clothes, to transport him from legendary tales of adventure on desert mesas to his new bustling urban home.

Or maybe he transported me. It is difficult to say with Coyote, a being drawn to boundaries because he enjoys mocking them, for the simple fact that lines are just lines, made to be crossed and flaunted.

In any case, you'll find him pawing around in the preludes to each section of the book, inhabiting a new variety of urban legend.

The city is the perfect place to find lines and boundaries and fences and all manner of rules about what goes where and what is and isn't allowed. So Coyote is here. He had to be. He couldn't help himself.

. . .

The city is a human-constructed membrane, a rearrangement of organic and inorganic materials. It bears affinities with a termite mound, a beehive, or a monk parakeet's colonial nest. Railways, highways, flyways—the exchange of food, disposal of waste, transference of energy, creation of technology—where does a city actually end? The urban reach encircles the globe. Yet a city is a cultural artifact. No more, no less. As such, a city will endure for only a time. The historical record shows that a city's existence *depends*—in the strict sense of that word—because a city is deeply dependent on the quality of the relationships that create and sustain it.[9]

Chicago marks the spot where a soporific river meets a massive, glacier-carved lake, oak savanna blends into tallgrass prairie, shrub swamp nuzzles against flatwoods, and humans encounter non-human otherness. A mere two hundred years ago, Chicago was a portage site, a way to carry a canoe from Lake Michigan to the interior waterways of the continent. Despite the massive changes imposed on the area since then, Chicago continues to provide common ground for diverse biota. In my explorations of Chicago, I've learned that cities are neither biological deserts nor derelict landscapes. They have their unique animals and their own *anima*.

Naturalists and writers are beginning to turn their thoughts and pens to the urban wilds, and my guess is that there are more to come. Same as me, people find themselves transplanted to cities with increasing frequency, including a new breed of ecologist that you'll meet in later chapters. These scientists are forgoing research projects in distant lands and exotic locales in order to study the lives of animals and plants that are entangled with our own.

By and large, however, popular nature writers still write about colossal landscapes, remote areas that test and refine the human will, and the gobsmacking glories and rawness of unpeopled places. Writing is often a solitary affair, both by necessity and because those who become writers tend to do well in keeping their own company, with full worlds inside their heads. Nature writers may go one step further. I was recently tickled to hear a nature writer complain of claustrophobia in her town of 3,400 souls in Canada. Another writer jokingly piled on that he and his partner, feeling similarly hemmed in, moved to the outskirts of their 150-person village. I don't blame them. As I said, that was my Plan A.

Yet we find ourselves where we are, some of it within our control, much of it not, and like Coyote—if we're smart enough to follow his lead—we adapt. These are the stories of my own adaptation to the city, the adaptation of other animals to the city, and how we might better adapt our cities to the larger landscapes on which they depend.

I

Inhabitation

Coyote Rolls the Dice

One day, Coyote was playing a game of dice with Badger and Wolf—and losing badly. "Ha!" barked Badger at Coyote, sweeping the dice up in his paw, "I win again." (Badger was not known for his magnanimity.) "It's just not your day," Wolf smirked at Coyote while casting a conspiratorial glance toward Badger.

The pair certainly seemed to be enjoying Coyote's loss, a feeling considerably sweetened by past grievances. They were frequently on the receiving end of Coyote's many tricks. Even when Coyote loses, Wolf thought, somehow you know you didn't win. But for now Wolf was satisfied, smug even. Badger was no better. Honestly, Badger was gloating. "Ha!" he repeated. He couldn't think of anything worth adding, so he said again, "Ha!"

The sun began to set, throwing a wash of shadow on the trio of gamblers. "Last roll and then we're done," huffed Wolf with authority. He handed the dice over to Coyote. With a pained expression on his face, Coyote assessed the situation. "Seems I'm in a spot of trouble, gentlemen," he murmured, eyes fixed on the dice in his paw. "I needn't remind you that the odds don't favor me. But as I've told you before," he said, brightening, "I don't worry myself with odds."

Before the last of those words tumbled from his mouth, Coyote tossed the dice across a bare patch of soil. The three companions intently watched as the dice rotated, thudding against the dirt, turning their way into what was certain to be a losing combination.

Then a strange thing happened. The dice struck the edge of an unyielding and mysterious barrier, abruptly skidding to a stop.

The three companions ran to the dice, eyeing them from above. "Cheater!" wailed Badger. Wolf shook his large head slowly.

"The dice can't lie, my friends," chuckled Coyote. "They fall as they may."

"But they didn't finish rolling!" Badger pleaded. "It doesn't count. Roll again. Do-over."

"Fair is fair," replied Coyote, a twinkle flashing like starlight in the corner of his eye.

"Who cares about this game, anyhow?" said Wolf, not bothering to mask his sourness. He pointed to the barrier. "I want to know what *this* is and how you put it here."

Coyote feigned innocence. "How should I know?" he said, sniffing at the flat, too-level surface.

For a moment, the game of dice was entirely forgotten. Each of the animals took a turn pawing at the odd, unyielding substance, scratching it with their claws, nibbling on it with their teeth, pushing at it with their heads.

"It doesn't budge," said Wolf, visibly perplexed.

"I can't dig through or under it at all," complained Badger.

Coyote paused, contemplating. Then cocking his head at his companions, he sprung theatrically into the air, landing with bravado on the hard, flattened surface. Wolf and Badger gasped in unison.

"You haven't seen pavement before?" he asked his incredulous associates. "Well, it seems to lead *somewhere*," he said, bounding up and down, delighted by the shocked expressions on his companions' faces.

"I don't like it," said Wolf.

"Not one bit," said Badger, finishing Wolf's sentence.

"You two sticks in the mud," clucked Coyote, "No fun. No spirit."

"I don't like it," Wolf repeated.

"Not one bit," said Badger, again finishing Wolf's sentence.

"Suit yourselves," replied Coyote. Trotting casually back to his companions, he snatched up the dice.

"Hey," snarled Badger, "those are mine." He slashed at Coyote, who easily leapt backward as Badger's sharp claws swished through the vacated air. Coyote stopped bouncing.

"Give Badger back his dice," Wolf growled.

Coyote held the dice out in front of them. "Well, certainly," he said, baring his teeth. "Come and get them."

Wolf and Badger made as though to move but couldn't bring themselves to cross the threshold where the mysterious not-soil began. The pair snarled and puffed out their chests.

Coyote looked upon them with pity. "Tsk, tsk, my friends," he said, pocketing the dice.

He pivoted, shook his tail as a final insult, and coolly loped down the concrete path—all without so much as a twist of his head back in his companions' direction.

Wolf and Badger stared with a mixture of astonishment, anger, even a dash of regret, but they could think of nothing else to do but return home.

The Channel Coyotes

The first sign was the rabbit bones.

I'm a wanderer by inclination and a walker by choice. When we moved to Chicago, our family of three gave up our car, a Honda Fit affectionately named the Trusty Grape for its dark purple hue (and as a contrast to our previous mechanical nightmare, the Silver Lemon). Life without the Grape is good. Not having to find parking or pay for anger management classes compensates for the small inconveniences of not owning a car. Another ancillary benefit is the gift of exploring the city by foot, at a pace conducive to free-ranging thoughts.

In the evening, I frequent the public golf course near my condominium. Not for golf, which is another gateway to anger management classes, but for sauntering. Off a neighborhood street, under the Metra rail tracks, over a rusty bridge that spans the North Shore Channel canal, I reach the fairway of the seventeenth hole. Flanked by silver maple trees and milkweed pods, the fairway provides enough of a long view to tap my inner *Watership Down*, allowing me to pretend I'm strolling through a meadow in the English countryside. This is where I discovered the rabbit bones. The following evening, they were gone.

I had a vague awareness that coyotes lived in the city. I can't remember who told me, but the news came as a welcome surprise rather than an unwelcome shock. In the first months after our move, we were renting a third-floor walk-up apartment with a window fifty feet from the elevated train (known colloquially as the "L") that runs to and from Chicago. Reading late one night, between the periodic squeals of the train, I heard a competing set of squeals, immediately identifiable as coyote banter. My ears prickled, followed by the hairs on the back of my neck. A visceral memory came to me: I once listened to similar yips and caterwauling surround me in full-stereo sound when camping in New Mexico—eerie peals of distressed laughter, redounding off high desert rock, prompting a smile and an increased pulse rate. In the apartment, I laid the book on my lap—for that moment, Chicago had become a New Mexican wilderness. I couldn't help but smile again. Welcome to Chicago.

Later I found out that urban coyote chatter like the kind I heard that evening is somewhat unusual. In Chicago, coyotes have adapted to city living in many ways—hunting at night, typically giving us as wide a berth as they can, and keeping their vocal hysterics to a minimum. Stan Gehrt, a scientist who has studied Chicago's coyotes for over twenty years, calls them "the ghosts of Chicago."

But they do materialize every once in a while. My first visual confirmation of a coyote—very likely the one responsible for the rabbit bones—occurred on another set of commuter train tracks near the golf course. Unlike the "L," the double-decker Metra train stays closer to the ground. The Metra's tracks bend outward for miles from their epicenter in downtown Chicago. One of those train lines shoots right through my neighborhood.

On a weekday morning, alongside a motley gathering of commuters, I await my eight o'clock ride to downtown Chicago. When we later detrain, the streets will be a chaos of bodies jostling without ever quite touching. But for now, we all mill about on the platform, closer to sleep than to work. The eyes of a besuited older man glaze over the *Tribune*. A woman brushes lint from her pant leg. A young executive grips his to-go coffee cup as though it's a lifeline.

That's when I see the coyote. He ambles across the double lines of railroad track, facing south, facing me, who faces him and says— too loudly, too giddily—"Coyote, coyote!" A lateral gravity draws me to the edge of the platform, where I stare in shock, even though the headlight of the train already bears down on us both. The coyote pauses for a quick check over his shoulder—perhaps setting his internal clock, having just wrapped up a night's work—and slips down the embankment into a narrow band of shrubs. A man in a neatly ironed shirt glances up, startled by my behavior. The expression on his face wordlessly poses a question: "Drinking before work, bub?" By the time he swivels his neck, following my transfixed gaze, the city's trickster, the wild canine ghost of Chicagoland, has melted away.

An accurate count of coyotes in Chicago is harder to find than coyotes themselves. I've heard rumors there are three thousand, possibly more. One thing is clear: the count is rising. Researchers removed twenty coyotes in 1989; in recent years, those figures have ranged from three hundred to four hundred relocations or euthanizations per year. That's a lot of coyotes. Hundreds now sport radio collars, so a good deal of information is known about their movements and territorial habits, even when they choose to be spirits unseen. What these electronic data points tell us is that coyotes are not just on the outermost edges of the city. They are in our city center. Coyotes are willingly moving to Chicago.

The reclamation of city habitat by wildlife is a national (and global) trend. A number of reasons account for the urban option: prey availability, human pressures on non-urban habitat, laws against hunting and trapping in metropolitan areas. In addition to these factors, coyotes are well suited for the task of city living—quintessential adapters, they consistently defy human expectations.

If one needs an expert on the comings and goings of Chicago's coyotes, Gehrt is the person to know.[1] Out of his office at Ohio State University, he leads the largest collaborative study of urban coyotes in the world. I attended a research symposium where Gehrt showed a video that sent a wave of chuckles through the room. Traffic, a bicyclist, empty fast-food joints—an ordinary Chicago intersection at night. Then, into the camera frame, a spry coyote trots casually to the edge of the sidewalk. She waits for the stoplight to turn red. Then crosses the intersection and continues on her way. Just another night on the town.

According to Gehrt, many Chicago coyotes have learned traffic patterns within their territories. As they patrol these territorial boundaries, the coyotes who survive are those who give due deference to the laws of the street. Available statistics reinforce that what I saw on the video was not an isolated incident. Auto naïveté doesn't serve one well in the city, two-legged and four-legged pedestrians alike, so although death by car is the leading cause of mortality for urban coyotes, it appears that many are learning and adapting to the unforgiving ways of the road.

Gehrt compares a coyote's perspective of the city to a patchwork quilt. The average territory for a coyote family group covers two square miles, with a large degree of variance depending on available food sources. The more concentrated the prey—which includes rodents, rabbits, deer fawns, goose eggs, and whatever else is available—the smaller the territory. The boundaries of these territories remain invisible to us, but through aromatic and vocal signals, coyotes know which lines they can safely cross.

In the field and at the podium, Gehrt speaks frankly about his awe regarding coyotes' abilities to adapt to urban environments. If it weren't for the radio collars that allow us to peek in on how they've managed to make a living in the most unlikely places, he is certain no one would believe it. They're that resourceful, that clever.

The philosopher of science Donna Haraway once wrote that our world is chock-full of confounding and witty "coyote discourse," which is "embodied in Southwest native American accounts [that] suggest the situation we are in when we give up mastery but keep searching for fidelity, knowing all the while that we will be hoodwinked."[2] Haraway is right: our lives are entangled with coyote discourse, and increasingly in the city with the discourse of *actual* coyotes, who remind us that wildness moves amid urban areas that we often think of as solely, or mostly, our domain. Coyotes let us know that the mental boundaries we keep—between the human and the wild—are more porous than we may have ever imagined. In the midst of our attempts to control the landscape, to put humans here and nature there, coyotes express an alternative set of ideas about boundaries. If we open our eyes, ears, and minds, a world full of coyote surprises awaits.

These notorious tricksters have escaped from their southwestern myths and go about their business largely undetected, in cities far from their familiar haunts in Taos and Tucson. Vancouver, Portland, San Francisco, Denver, New York City—all now have thriving coyote research programs. Occasionally, a coyote makes a bold appearance. A few years ago, on a steamy summer day, one found his way into a Quiznos' drink cooler in downtown Chicago. Such brazen acts are rare, yet more people are finding out what is known among coyotes: wildness is part of the city, if you know how to look for it. Keep your eyes and ears open. Or watch the train tracks.

Or head to a golf course at dusk. This past summer, while walking my son home from school, he and I noticed several nighthawks in the neighborhood, sweeping the air for insects. Nighthawks perform aerial acrobatics that would be well worth buying tickets for; they can slice and twist with the precision of top-gun pilots. Feeding on the wing at high speeds, they flash their white-barred underwings with mouths open wide, catching moths, gnats, flies, and other flying insects in their gaping maws. My son and I decided to take a detour to the golf course by the Channel canal, hoping the combination of insect fecundity and open space would lure the avian hunters.

What a show it was. After arriving, we got low to the ground and army-crawled on our bellies to the edge of the grass, eyes darting between birds, mouths as agape as those of the nighthawks, who were reveling in the feast on the wing. The finale, however, came courtesy of a surprise visitor. As we lay motionless, watching the birds, a flash at ground level drew our attention: one of the ghosts of Chicago. A Channel coyote sprinted across the open fairway at a full trot, returning from his own neighborhood adventures. That evening made my son's top-ten best days of all time.

A chain-link fence runs most of the length of the Channel canal, keeping people from the water, so if you're not willing to clamber up and over, you can get to the canal only at special access points. You can, however, see the vegetation on the Channel's embankments from the bridges and roads that cross over the waters. When my son and I left the show that night, heading back home, we looked down and spied the hollowed-out base of a large cottonwood tree. Difficult as it was to see through evening's gloam, we thought we could discern— just maybe—the contours of a coyote head staring back at us. My son was positively sure. Coyotes can play tricks on the mind like that.

I take a lot of walks on the golf course by the Channel. Sometimes I see the coyotes who live there; most times I don't. Once around sundown, a flash of rufous fur vanished in the grasses just as I was turning my head. It could have been a dog—but I know what ghosts look like now.

On another walk in a gentle rain, maple and linden leaves gathering the last daylight, shadows fingering their way across the grass, I see a pair of Channel coyotes—presumably a mother and father— and come to a dead halt. They confer a few minutes as I edge closer, finally vanishing into the underbrush. I walk on, unable to shake the feeling of the encounter and the mutual gaze we exchanged as they took my measure. I check one last time over my shoulder before I reach a neighborhood street. Behind me, fifty yards away, one of the Channel coyotes leans back on his haunches, watching me.

In *Make Prayers to the Raven*, Richard Nelson's ethnography of the Koyukon people of Alaska, he notes that the Koyukon believe the forest watches them to ensure that they behave respectfully toward the land and its inhabitants. According to Nelson, the Koyukon live "in a forest full of eyes."[3] Ever since I met the Channel coyotes, the city has become more *animated* for me—in the root sense of the word, *anima*: to possess breath, spirit, or soul. The Koyukon people live in a forest full of eyes. I live in a city full of eyes. Some of those eyes belong to coyotes, who are watching to see whether I learn how to adapt to a wild city.

Many Native American tales about Coyote, the trickster figure, begin with a standard opening: "Coyote was going along . . ." This phrasing bears some resemblance to the fairy-tale salvo "Once upon a time . . ." There are differences worth noting, though. "Once upon a time" locates a story in some distant world, refracted through the prism of fantasy. "Coyote was going along" suggests something else. The audience to these narratives is dropped on the trail with this character, in midstride. The events of the tale may have occurred long ago, in a distant time, but its protagonist keeps going, trotting through our present. Coyote is always up to something, always on the move.

In these stories, Coyote often finds his way into trouble, sometimes loses a limb, his intestines or anus, and occasionally he even dies. But death is not the end of Coyote. After all, he must keep *going along*. There will be a need for another Coyote story on another winter night.[4] So, when Coyote does meet a comical or tragic end, there's usually another animal on hand—Hare or Fox are popular resurrection specialists, depending on the tale's region of origin—to perform a ritual, step across his body five times, and put him back together. Sometimes, Coyote possesses magic enough to put himself back together.[5] He keeps going along.

I wonder if these tales reflect observations of actual coyotes' tendency to defy death, to keep *going along*. Despite their severe persecution in the United States, coyotes are thriving. The numbers of coyotes killed—and the potent chemical concoctions developed to control their numbers—is shocking.[6] More shocking still is the defiance of coyotes in the face of such persecution. Despite organized and well-funded eradication efforts, coyotes have overcome again and again, spreading across the country into the niches left vacated by gray wolves, their less-fortunate canine cousins. This uncanny ability to survive has much to do with coyotes' gifts as a species, what biologists call *behavioral plasticity*, meaning they are adept at altering their behaviors to fit their environmental contexts. In practice, this might mean grouping together or going solo, adjusting their omnivorous diets according to opportunity or cultural learning, and, most important, producing larger litters of pups in response to perceived population declines.

Try to kill off Coyote in a story and he'll pop back up again, *going along* to another adventure; try to kill coyotes, and they'll thwart your best efforts. The trickster Coyote's real-life descendants now populate nearly every city in the United States. After being driven away, purportedly cleaned out, harassed, subjected to chemical warfare, demonized, and half-forgotten by urban dwellers, they've returned. I get the sense that they'll keep finding their way among us, almost as if they are on a mission to jog our memories. This city, this land, is not exclusively yours, they seem to say; we'll keep coming back until you get it.

It's been a while since I've seen the Channel coyotes. I began to think they had moved on to more attractive urban hunting territory or that some misfortune had befallen them—possibly those deadly automobiles caught them in a moment of distraction, when they were preoccupied with a rabbit nibbling grass in someone's front yard. Then, as I was cutting across the golf course late one night, I saw a canine silhouette through the darkness. Only for a moment. Still going along.

I walk on, immersed within the city like the Channel coyotes, adopting it as my own, scratching out a place in abundance, exploring the margins for life and further connection. As an urban transplant, I sometimes feel a long way from home and a long way from my ancestors. When the city presses in upon me, coyotes remind me of the vitality that weaves its way between the buildings. Humans may often disregard, displace, and disrupt other kinds of animal life, but the *anima* of what we now call Chicago is not gone. The coyotes keep it flowing; they keep going along, beckoning us toward greater fidelity with our nonhuman kin. Lead on, coyotes. Show what a city can be.

Scrapers of Sky

> Know this Primal Power
> that guides without forcing
> that serves without seeking
> that brings forth and sustains life
> yet does not own or possess it
> —TAO TE CHING, verse 10

Skyscrapers and canyon cliffs are not typically associated with each other. We are trained to think of wild, river-sculpted landscapes as sites where human traces remain only as faint boot prints or the scattered ashes of a campfire. Habitat is habitat, though. Peregrine falcons, who have begun to nest in and hunt the accidental canyons of the city, seem to know this.

A Chicagoan doesn't need carabiners, a harness, and climbing ropes to see peregrines *in situ*. The ability to ascend a flight of stairs will do. My local public library hosts a peregrine pair who have repurposed a cement nook beside a third-floor window as their aerie. During the spring and early summer, a live "falcon cam" broadcasts the parents' daily fussing over their eggs, and later, when hatched, their fledglings. Affectionately named Nona and Squawker, the peregrine couple returned to raise a new set of chicks for the thirteenth time this year.

Nona and Squawker are one power couple among many in the Chicagoland area. Across Illinois, more than two dozen peregrine pairs are monitored by the Field Museum, and the majority of their aerie sites are in Chicago and its suburbs. Peregrine nests are in the South Loop, Waukegan, Calumet, Millennium Park, on Wacker Drive, and at the Uptown Theater, among other urban locations. They've found Chicago homey enough.

For other city animals, peregrines are cause for concern. There goes the neighborhood. A peregrine attack is called a "stoop," a dive-bomb technique that the raptor uses to surprise other birds from above, stunning or tearing at targets that can be the size of a sandhill crane. Incongruous as it may seem to have a bird in the city that can reach speeds of over two hundred miles per hour as he or she drops toward a prey animal—three times faster than a cheetah's top speed—this wild aviator is in our midst.

I'm an amateur birder at best, appreciative of serendipitous moments with the avian world. I'm an onlooker more than a life lister. Birds of prey, however, hold a particular fascination for me. When I found out that peregrines were nesting in the same place I fail to return books in a timely manner, I got a twofer: I paid the raptors a visit and then made use of my library card, turning to a classic source for understanding peregrines' daily lives, J. A. Baker's natural history classic *The Peregrine*.

Set in coastal southeastern England and published in 1967, Baker's book provides a remarkable literary account of one man's obsession with peregrines. Early on, Baker observes that the differences in British landscapes are "subtle, coloured by love." His writing style takes on the affects of the birds he so closely observes: mostly spare, few wasted strokes, each word alive with precision and calculated beauty, at times sweeping over the landscape and then ranging upward in transcendent spirals. The meat of the book—where the reader follows Baker as he becomes more and more intimate with the nuances of the birds' behavior as well as the shifting moods of the land—induces a hypnotic effect. Baker unfurls a poetics of place, coloring everyday scenes with extraordinary brilliance.

Baker's eye for the individuating detail is honed by his interactions with the peregrines themselves. As he observes: "Hawk-hunting sharpens vision. Pouring away behind the moving bird, the land flows out for the eye in deltas of piercing colour. The angled eye strikes through the surface dross as the obliqued axe cuts to the heart of a tree. A vivid sense of place grows like another limb." As he comes to understand the personality quirks of particular peregrines, Baker's prose absorbs the birds' sense of immediacy: "What is, is now, must have the quivering intensity of an arrow thudding into a tree. Yesterday is dim and monochrome. A week ago you were not born. Persist, endure, follow, watch." The narrative also pays tribute to England's coastal landscape—its subtle tonal shifts, violence, and ceaseless movement of color, light, and shadow. Wandering through woodlands and farm-studded estuaries over the span of only a few winter months, Baker offers a keen-eyed perspective applicable to any landscape at any time, including cities such as Chicago, where peregrines increasingly fledge their young and cleave the air with their wings: "I have tried to preserve a unity, binding together the bird, the watcher, and the place that holds them both."[1] Could we learn to see a city as a peregrine does? Would this perspective bind us to a new vision of what a city can be?

Baker wrote *The Peregrine* in the late 1960s. At the time, peregrines in England were winking out, their nervous systems fatally compromised by pesticides, their DDT-thinned eggshells too weak to hold a baby bird to term. In the United States, once denial turned to alarm about DDT's pervasive impacts, it was banned in 1972. Around the same time, the Endangered Species Act solidified critical federal protections and sparked a desire for peregrine recovery. In the 1980s, researchers began to raise peregrines for release. The results of these efforts can be seen streaking between the buildings: peregrines are in the city's canyons, on the hunt again.

"Oooo! Oooo! Oooo!" My colleague Anja Claus reminds me when I forget. "Oooo! Oooo! Oooo!" has become her way of alerting work-mates to the thrilling fact that a peregrine is circling. On the twenty-eighth floor of the Civic Opera Building, we have an excellent view of the South Branch of the Chicago River as well as the spread of the city's canyons to the west-northwest. We hurry to the windows, cran-ing our necks skyward for the distinctive silhouette, the boomerang of wings with a bend in the middle, the tapering wedge of tail feathers. "Oooo! Oooo! Oooo!" is the appropriate response to a creature who arcs between buildings with such quicksilver grace.

The scientist in Chicago who may know more about these wild birds than anyone else is Mary Hennen, a collections assistant at Chicago's Field Museum and a self-described "liaison" between per-egrines and the many volunteer bird monitors who have taken an interest in them. In addition to managing the Field Museum's bird "library"—a collection with close to half a million specimens rep-resenting around 90 percent of the species in the world—Mary has been doing peregrine research for almost twenty-five years. Orni-thology, however, wasn't what she had her sights set on in 1987, when she was fresh out of school with a biology degree from the University of Wisconsin–Stevens Point. Her volunteer work at both the Chicago Academy of Sciences and, soon after, the Field Museum led her into the enthusiastic orbit of a couple of top-notch ornithologists. Mary describes her good fortune as a matter of being in the right place at the right time. Her budding scientific research career intersected with the first peregrine releases in Illinois, an effort to recover per-egrine populations in a region that, as in all areas east of the Rockies, lacked peregrines by the 1960s. Before reintroductions, the last per-egrine nest in Illinois was recorded in 1951.

When the first releases occurred, the team at the Chicago Acad-emy of Sciences had modest hopes: three breeding pairs in the state. There are now close to thirty. Most of the peregrines in Illinois main-tain year-round territories, but some have lived up to their scientific name (*Falco peregrinus*, Latin for "wandering falcon"), dispersing as far as Ecuador and Venezuela. In 1999, peregrines were delisted from the national endangered and threatened species lists; in Illinois, the birds dropped from endangered to threatened status in 2004, and in 2015 they were delisted. By any measure, this is a high-profile species success story, and Mary has been instrumental in monitoring and assisting in the peregrines' progress.

Mary's work comes with hazards. As we speak, she reaches underneath her desk and retrieves a bicycle helmet. The helmet, she explains, keeps the crown of her head buffered from the taloned fists of aggressive peregrine mothers and fathers, who are not nearly as interested as researchers in the scientific benefits of having their babies banded for monitoring purposes. Mary has adapted. Because peregrines are drawn to the highest point of an uninvited guest, a whiskbroom held aloft by a partner is one way of keeping her head free of peregrine-induced lumps, as well as keeping the birds free of injury.

Mary shies from taking credit for the success of peregrines in this area. She is particularly sensitive to instances when people describe peregrine releases with words such as *place* or *put*. The agency of the birds, their wildness, is a quality she holds in high esteem. "They are wild birds, they go where they want," she explains. "It's the birds finding places to breed on their own and doing that successfully."

Other than the wild tenacity of the birds, a key factor for peregrines' success may be the city itself. Peregrines are historically cliff-dwelling raptors, and as Mary observes, "If you think of the city, it's nothing but a pseudo-cliff, with lots of ledges, ample prey, and no competition for use of the space."

Conservation scientists sometime refer to human-built environments and structures that other species utilize as "habitat analogues." Peregrines reveal how such analogues can be critical to rare or threatened species. In plain speak, humans build stuff that other animals will use. The presence of peregrines in Chicago sheds light on a larger phenomenon—and perhaps an important new mindset—called *reconciliation ecology*.

Restoration ecology, in fits and starts, has become a major piece of conservation practice. Chicago is an epicenter of citizen-led restoration projects, which can be traced back to early attempts in the late 1960s and early 1970s to re-create or expand the habitat of conservative prairie flora. Restoration ecology, defined succinctly, is the (often sweat-intensive) process of bringing a historical ecosystem or landscape back to a condition resembling its former functionality and diversity.[2]

Reconciliation ecology—a term coined by evolutionary ecologist Michael Rosenzweig—"is the science of inventing, establishing, and maintaining new habitats to conserve species diversity in places where people live, work, or play."[3] In his book *Win-Win Ecology*, Rosenzweig offers up several case studies to consider, situations in which people both intentionally and unintentionally have created critical habitat for other species while still making a living themselves: ongoing red-cockaded woodpecker habitat management on an active air force base in northern Florida; Chiricahua leopard frog recovery conducted by a coalition of ranchers in the American Southwest; and a reconstructed pseudo–salt marsh that is a boon for migrating birds in Eilat, Israel. All share a common thread: with the proper adjustments and adaptations, humans can live—and make a living—alongside other species, and both can thrive. So while restoration ecology involves diminishing human impact in particular places so that other beings can thrive or reestablish themselves, reconciliation ecology advances the position that humans can create and build novel systems that are suited to other species. In short, by understanding the behaviors of other species and what they require to meet their needs, we can deliberately create places of cohabitation.

Reconciliation ecology will work for some species and not others, so conservation still needs restoration and preservation in its tool kit to accommodate different species. Many rare and threatened grassland birds in the greater Chicago region, for instance, cannot survive without large contiguous natural areas for their breeding and nesting sites. Rosenzweig, however, has strong words for conservationists whom he believes have misdirected their energies. By focusing on setting aside unpopulated acreages as a last hope, he argues, conservation efforts tend to neglect areas that could put us into daily contact with wild animals. Maybe the preservationist strategy was an appropriate response at one time, but the world has grown smaller as the human population has grown larger. Large protected nature reserves remain crucially important to some species, but as Rosenzweig puts it: "We must abandon any expectation that reserves by themselves, whether pristine or restored, will do much more than collect crumbs. They are the 5 percent. We need to work on the 95 percent."[4]

The city makes up part of this 95 percent and presents a trove of possibility. As Rosenzweig's research attests, we may be surprised at what kind of wild animals will live among us or in close proximity to us, if given the opportunity. In this respect, perhaps peregrine falcons should be the poster animal for reconciliation ecology in the many city skies they've reclaimed. From Nottingham, England, to Hong Kong, peregrines are on the hunt.

A short walk from where Mary Hennen, the peregrine expert, works at the Field Museum, there is a piece of land that extends into Lake Michigan called Northerly Island. When I am anywhere near the eastern edge of downtown Chicago, I make an effort to go there, seeking refuge for my senses. Not coincidentally, so do other animals. Northerly Island might be one of the more striking examples of reconciliation ecology in the heart of the city.

The island, which is actually a human-made peninsula, reminds me that the city is a place of constant change, and with the right intention that change can benefit both humans and other species. Northerly contains several dozen acres of rolling prairie, thoughtfully planted and beautifully diverse, and an unbeatable view of the liquid expanse of Lake Michigan to the east. To the west, across an hourglass-shaped harbor, one can see Soldier Field (and hear the ocean roar of fans when the Bears are playing at home on Sunday). To the northwest is Chicago's downtown skyline.

One hundred years ago, there was no island here at all, only open water. Northerly represents the fruits of lakeshore planning and beautification inspired by the visionary architect Daniel Burnham. Visionary plans don't always make it past the drafting table, however.

The sole portion of the six-mile parkway Burnham proposed that was actually built was Northerly Island, so named because it was the northernmost point in that grand plan.

If you stood on this spot in 1933, you would have been shouldering your way between crowds of people and architecturally daring buildings, celebrating the second World's Fair hosted by Chicago (the Century of Progress International Exposition). By 1948, you would have been dodging small aircraft, for Northerly Island had become Meigs Field, a single runway airport that catered to business-people, commuters, and state politicians. So it was for over fifty years. Sometime in the middle of the night on March 30, 2003, however, you would have been avoiding bulldozers as they tore X's into the air-strip, signaling the demise of Meigs Field. In a controversial move, Mayor Richard M. Daley chose to forgo further negotiations about the airport lease and reclaimed the island on behalf of the city for its original purpose as parkland.

Because it is situated along a major bird migration pathway, Northerly offers a welcoming site for long-distance flyers who hug the coast of Illinois as they follow the Lake Michigan shoreline. Though modestly sized, the patch of land provides significant habitat for rest and refueling, now for birds instead of airplanes. Some avian visitors linger for longer stretches of time. Northerly is one of the few places near downtown Chicago where you might hear the song of a grassland denizen such as a dickcissel or see the flash of a kingbird's tail as she picks off insects between tree branches. Other visitors one wouldn't expect in Chicago—including short-eared owls, northern shrikes, horned grebes, bluebirds, and even an occasional snowy owl down for a visit from the Arctic—are known to make stops at the is-land as well. I've spotted a muskrat paddling the five-acre lagoon at the island's center, and, considering the gaggles of Canada geese who regularly promenade along the island pathways, Northerly must make an alluring coyote hunting ground.

A visit to this downtown island provides an echo of what portions of Chicago looked, felt, and smelled like before there was a Northerly Island, before there was even a Chicago. Northerly suggests the many ways that a city can accommodate itself to other species. Some animals—like peregrines, chimney swifts, and cliff swallows—adapt well to the structures we build. Some animals—like coyotes, raccoons, and opossums, if they can stay out of harm's way—do a fairly good job of making use of what is on offer at ground level. Places such as Northerly Island provide for those animals who cannot tolerate a great deal of human disturbance yet are still willing to venture within close proximity of the city if the conditions are right.

The juxtaposition of skyscrapers to prairie, separated by a thin slice of water, indicates different forms of control and letting go. The peregrines, once they were released from the deadly grip of DDT, adapted to our built environment. Northerly takes reconciliation a step further, offering a thriving prairie savanna where once concrete and airplanes dominated.

Reconciliation ecology. I feel we are capable of it when I stand on the shores of Northerly Island, gazing at the Chicago skyline, hoping to see a crescent-shaped bird speeding through skyscrapers in the distance.

There is a good deal to learn from reconciling our needs with the needs of other species. Small-scale instances of reconciliation occur every day in school gardens, in backyards, on green rooftops, and atop abandoned "L" platforms. Reconciliation can happen at larger scales through the ongoing work to reclaim brownfields, creatively alter postindustrial sites, and connect green infrastructure and riparian habitats through the heart of the city.

When other animals are in our midst, their lives mean more than something in a textbook, a children's story, or a *National Geographic* special. Active reconciliation means trying to understand what enables other animals to flourish in our presence, and how we can proactively create such places of cohabitation. This can foster ecological empathy, opening up a space for the long-term work of living with grace and skill in our everyday worlds. Reconciliation ecology asks of us that we anticipate the impacts of our actions and take responsibility for our historical shortsightedness. Maybe it also asks that we say "Oooo" when we witness a miracle in the sky.

Every time I walk by the Evanston Public Library, I look up at the third floor, just in case. On rare occasions, I've been rewarded by the sight of Nona—or maybe it's Squawker—gripping a corner of the building, framed by blue sky. A slate-gray head with the distinctive black "mustache" streaked across the bird's cheeks. A deep pool of eye, rimmed by bright yellow, returning my gaze. People hurry past me, not looking up, not even looking at the odd man on the sidewalk looking up.

In *The Peregrine*, Baker reminds his readers, "The hardest thing of all to see is what is really there." Peregrines are here, although many people may be unaware of their return to the city's canyons. Made to cut the air, they circle above downtown streets, offering a glimpse of the changing ecology of cities—a reconciliation of our habitat and theirs.

Under Construction

The second-largest-bodied rodents in the world have arrived where I live. For those who don't keep a chart of rodent size rankings handy, I refer here to *Castor canadensis*, commonly known as the North American beaver. This is a creature whose plant-based diet can produce sixty or more pounds of plush girth.

Fur and rotundity can do wonders for an animal's public reputation. Beavers, along with squirrels, are one of the few rodents that most people willingly tolerate in close proximity. Many would call them cute. I unrepentantly count myself among those people.

The fur-bearing newcomers possess more intrigue than the cute factor alone. They've made a bold urban turn, swimming against the stream, so to speak. I first noticed their signature on the city while walking alongside the North Shore Channel, not too far from where I have encountered coyotes. Fresh tree shavings and tooth marks revealed a recent dining experience, although no beavers were out and about. Then, one day, as I crossed a busy street above the Channel, a strange wake in the water caught my eye. At the leading edge of a V-shaped ripple was a furry tuft of head, surveying the banks, possibly for dinner.

The uniform straightness of the Channel makes its waters hard to mistake as a naturally occurring river. Cut between 1907 and 1910, its purpose was "plain and simple, to get rid of the decomposing and noxious filth in the Chicago River's North Branch."[1] The Channel's quality is significantly improved since its open-sewer days, and I frequently kayak and canoe the waters in order to be transported, in body and imagination, into the urban wilds. I am heartened that many other animals are repurposing the Channel for non-sewage-related experiences, too.

Beavers are the latest creature seeking to further the Channel's re-development. It seems fitting that two kinds of construction have been united—an artificial canal, originally (and still) intended to deposit citizens' unwanted effluent into the Chicago River, and the incipient foundations of a beaver outpost. The laborers who built the North Shore Channel over a hundred years ago certainly didn't have beaver habitat in mind when they engaged the ditchdiggers, yet the beavers have accepted the unintentional offering with gusto.

Beaver resourcefulness is worthy of admiration. The colonial-era fur trade devastated their populations across the country. In the Chicago region, beavers were likely cleared out by the 1820s, and they were ex-tirpated from the entirety of Illinois by the early twentieth century. But their story wasn't over. Following a series of government-led reintro-ductions in various parts of the state, beginning in the late 1920s and continuing into the 1930s, they have staged an epic comeback. In the ab-sence of a public desperate for the latest in fashionable rodent hats, these toothy fur balls now can be found in every single county of the state.

And they can be found, sometimes unexpectedly, in the most densely populated portions of Chicago. An acquaintance told me about a group of beavers who furtively crisscross the parking lot of the Holiday Inn Mart Plaza in downtown Chicago, ever on the prowl for edibles. In the act of foraging for late-night snacks of shrubbery, these beavers have become known for surprising the car keys out of the hands of unsuspecting hotel guests.

. . .

Downtown Chicago or elsewhere, my first inclination is to welcome beavers anywhere they can swim or waddle to on their own. Granted, I don't have a favorite cottonwood tree sitting streamside—which is a delicacy in the school of beaver culinary arts—or a farm to protect from becoming a wetland, or a need for a startle-free parking lot. My stake in the matter is solely based on the thrill of having another nonhuman neighbor, and what that might portend for urban wildlife diversity.

Given beavers' long absence from the lives of most Illinoisans, particularly those of us in the city, their modest return to places in which they once thrived perhaps can be reduced to a single word: hope. Poet Emily Dickinson offers the soulful suggestion that hope "is the thing with feathers." Hope may also be a thing with a flattened tail that can crack the water with the force of a canoe paddle. Maybe we're on our way to appreciating and championing biodiversity, I tell myself. Maybe our cities are becoming more natural, or naturalized. Maybe the beaver is a package of living aquatic, fur-ball evidence that humans can willingly reside in the midst of a greater variety of animal life than previous generations of city folk.

Maybe we recognize something all too familiar in the industriousness of beavers. Both humans and beavers leave prominent marks on land and water. Both can dramatically change the ecologies of the places they choose to call home. Beavers come in second only to humans in the extent of total habitat altered (although the distance between first and second place is immense and perhaps not an achievement worthy of bragging rights).

Beavers' steady work ethic often creates wetland habitat for an entire suite of other species. Conservation biologists refer to them as ecosystem engineers, and the benefits of their exertions are well documented in areas to which they are native. I first learned about beavers in this context—for their positive roles in creating dynamic ecologies and opportunities for biodiversity richness. But that kind of scenario more accurately describes beavers in, say, Yellowstone National Park, where a few (or a lot) of downed aspen and willow trees have time and opportunity to regenerate after the beavers inevitably move on to new food sources.

What about beavers in urban areas? I'm not a biologist, but I had a hunch that any animal with the potential to change habitat so dramatically might run into difficulties in a city. So I called Steve Sullivan, urban ecology curator at the Peggy Notebaert Nature Museum, who offered a more sobering perspective about welcoming beavers with open arms and urban ponds. Steve told me there's been a bit of back-and-forth between urban land managers and beaver aficionados in Chicago. One contentious spat revolves around a series of beavers that set their minds on making North Pond a dinner table.

North Pond is a thirteen-acre oasis in the Lincoln Park neighborhood that hosts turtles, waterfowl, butterflies, and migratory birds, among other dependents—all watched over by mature trees, including revered, long-lived willows. You may be able to discern where this is going. Like the North Shore Channel, Chicago's North Pond is entirely a human construction and yet has become critical to a great deal of urban wildlife. If the beavers stay, then the willows will fall, because beavers, as do we all, gotta eat. If the scrumptious yet life-giving trees go down, the result would be an urban pond denuded of its key vegetation and therefore many of the key species that depend on this vegetation. In other words, instead of creating dynamic, vibrant habitat over the long term, the beavers would reduce the regenerative capacity of this urban ecosystem, decreasing overall biodiversity, then move along once the edible stuff is gone.

The beaver problem at North Pond suggests the larger problem of appetite that created this conflict of interests. Humans are unambiguously the number-one ecosystem engineers. To gain such status, we've initiated a great many landscape-wide changes, and the impacts of these actions upon other species is too often an afterthought, if it has been thought of at all. Other species become the casualties of this appropriation of stream and forest, pond and field, soil below and air above. The smack of a beaver tail at North Pond raises a question too rarely considered: how can humans be ecosystem engineers worthy of the *ecosystem* and not simply the *engineer*?

Whether one values beavers out of a sense of moral obligation to other animals, or from the reasoning that they once lived here and should be allowed the freedom to do so again, or from simple delight in seeing beavers be beavers—I don't fault anyone for thinking that beavers deserve an opportunity to exist, even in the heart of the city. As with other animals who are becoming our urban neighbors, I hope this is the beginning, not the end, of a conversation about creative coexistence and our shared responsibilities to our waterways. When it comes to an urban water ethic, there is much construction to be done.

■ ■ ■

I can smell change in the air today. Wood smoke wafting from chimneys. An electric current of cold in my nostrils. I am instinctively drawn to the Channel waters on days like this. Beavers are on my mind. I have come for a closer, more intensive look at the place where, many months prior, I'd first spotted scattered piles of bark shavings that clearly showed that beavers were in the neighborhood. I find a lot of beaver signs: small linden trees toppled and lying prone; the bases of large ashes gnawed halfway through; a stump, two-feet in diameter—with telltale incisor hieroglyphics—coning upward to a ragged point. An impressive de-construction site. I also discover four separate beaver lodge openings dug into the embankment.

Yet something seems awry. Three openings are covered with the remnants of fall, a patina of faded brown maple leaves and broken twigs. I get down on hands and knees to get a closer look, a few footsteps away from the languid Channel waters. This hole is the only one that appears as though it still might be in use. Seeing no evidence of beaver tracks in the slippery soil, I tilt my head upside down between tree roots, peering as far as I can down the dark passageway. My mind conjures a vision of an angry beaver lumbering out and spilling me into the water like an outtake from a screwball comedy. Fortunately for me, order prevails.

With the knowledge that the city's animal warden might have reasons to cast a disapproving eye on beaver loitering, I wonder if what I am observing constitutes a crime scene. Perhaps the aged beaver chews, as well as the lack of discernible tracks or fresh leavings, contain clues that point toward an abduction. If so, there would be no beaver profiles on milk cartons and no 911 calls.

The explanation for the absence may be simpler: abandonment rather than abduction. Sensing better prospects on the urban frontier, the beavers could have pulled up stakes and moved up the Channel to tastier meals. Or it is possible that the beavers are currently holed up and prepared for winter, with a fully stocked larder of foodstuffs, awaiting the coming spring greens.

One last possibility: they could have begun the journey down the Channel, following the gentle current to where its waters join the North Branch of the Chicago River. From there, they might have swum still farther to reunite with second cousins once-removed in Chicago—their relatives who are currently startling hotel patrons who only want to get to their cars without incident.

A month or so later, after winter had begun in earnest, the mystery resolves itself. Near the spot where I conducted my amateur sleuthing, I pick up the trail of beaver presence. I put my head down and follow the equivalent of beaver breadcrumbs—golden sapwood shavings and decapitated heartwood—to a cluster of downed trees. One of these newly fallen masts sprawls down the bank, reaching its topmost limbs across the edge of the Channel waters. Apparently the carpenter's shop was still open for business.

Gusts of icy snow dance through spindly branches, pricking at my cheeks. I can see through dark, unsleeved tree limbs to the iced-over waters of the Channel. There, where the downed tree lies, is a four-foot opening in the ice. And there, on top of a branch, nibbling like it is his only task in all the world, is a beaver. Not wanting to startle him, I get as close as I dare, maybe twenty feet away. I resist the cold as best as I can, then lose time watching the furry engineer—shifter of currents, creator of wetlands, bane of urban ponds. His head is covered in a dusting of snow, the fur around his face pulled into points, as though he's just toweled off post-shower. He twists a small stick between his chocolate-colored hands, gnawing the bark off as he rotates it, jaws working feverishly. The bead of his black eye appears unfocused, barely registering the snow as it falls while he savors his winter morsel.

I watch his deft fingers turn the stick and think about the construction of cities, how even when we build for our purposes alone, other animals may find an offering in disguise. Where we see waste, they may see opportunity. Beavers seem to possess a wisdom for ecosystem engineering that supersedes our own. When unconstrained by pavement, they are change makers with a mind and a set of incisors that creates abundance for others while meeting their own needs. As they reoccupy urban waterways, they may carry a hefty message along with those sixty-pound bodies: a good engineering project should be measured by the quality of relationships it creates. I hope beaver intentions and human attention can be united to build a more habitable city together.

The snow continues to fall, hushing the landscape. The cold begins to seep through my layers of clothing, and I need to move. The beaver keeps working at the stick. I stifle the urge to yell, "Welcome!" into the winter air. I whisper it instead, then turn to go home, back to my lodge.

The TV Graveyard near Tong's Tiki Hut

We are at the midpoint of our journey. Seth twists the key, turning the truck ignition off, and we step out and walk the few remaining feet to Tong's Tiki Hut. The aroma of cooked rice wafts toward us. We take our seats underneath yellowed mood lighting. Baskets of faded palm fronds and plastic gulls dangle from a ceiling covered in fishnet cordage. A brightly colored totem pole in the corner confronts us, three stacked heads summoning up monsters from otherworldly shores. The faces gaze, eyes wide, issuing a warning: this is not your island. My attention drifts toward a wall painted with a crystalline lagoon, encircled by a crescent of peach sand. A humming air conditioner creates a slight breeze.

We are not in Fiji. We are west of Chicago. We are on an adventure of sorts. How exotic is up for debate. But it cannot be denied that we are eating lunch under palm fronds and a large fish net.

My companion is Seth Magle, a wildlife ecologist at Lincoln Park Zoo's Urban Wildlife Institute (UWI). Founded in 2008, UWI has accomplished a great deal in a short period of time. Seth is the second director, not far removed from a doctorate that had him studying black-tailed prairie dog colonies in suburban Denver. When he took the position at the zoo, he quickly got to work on an innovative and long-term research project that aims to monitor and provide a comprehensive inventory of urban wildlife in Chicago. "We all live in an ecosystem; we just don't know it," Seth tells me as he scans the noodle entrées. "When it comes to urban areas," he adds, scratching at his sand-colored beard, "people have an ecological blind spot."

Seth and his colleagues are trying to eliminate that blind spot. The Urban Biodiversity Monitoring Project, which he leads, is the largest and most systematic attempt to collect information about urban wildlife in the world. The project relies on approximately 120 motion-triggered cameras along multiple transects that radiate outward in all directions from the city's core. The cameras are mostly affixed to trees—beneath underpasses, in pocket parks, next to parking lots, or anywhere a good angle can be found to spy on a roaming city critter. Gathering these photographs enables Seth and his team to build predictive maps based on animals' behaviors. I've seen some of the pictures: skunks, opossums, white-tailed deer, coyotes, raccoons, foxes, cottontail rabbits, beavers, even the occasional mink or flying squirrel.[1] There is also the occasional two-legged critter. Despite durable casings and camera locks, Seth has lost a few cameras to these two-leggeds. The price of science.

We finish up at Tong's and get back in the truck. I crack open a fortune cookie, ignoring the culinary inconsistency with Polynesian cuisine. "Travel is in your future." *Nailed it*, I think, crunching the cookie between my teeth and looking out the passenger window. We get back on Roosevelt Road. I'm not sure why, but a quip by Henry David Thoreau arrives unbidden in my mind: "Eastward I go only by force; but westward I go free." In Chicago, going east means eventually going into Lake Michigan. Today, we travel due west.

Our task involves stopping at various points along the transect to download the images that the cameras have captured onto a laptop computer. Seth opens a camera box attached to a tree on an embankment below some Metra tracks. This is when he mentions that he sometimes feels like an exile. Not from his home country, but from his home discipline of wildlife ecology. He has friends in Tanzania, the Congo, Kenya, Australia. When they find one another at professional gatherings, he is often regaled with tales of close calls involving raging hippopotamuses or near-death experiences in malarial rain forests. Seth's stories involve avoiding drug dealers on his way to fetch camera data in urban parks or finding a place to relieve himself near an interstate highway.

This is to say that Seth represents a new subspecies of ecologist, and there isn't much geographical precedent for his work. Since the beginning of the discipline, the textbook heroes who developed evolutionary theory and launched the field of ecology did so on the basis of their discoveries in foreign locales. As a twenty-two-year-old who almost became a country parson but instead opted for exploring the mysteries of creation, Charles Darwin famously sailed to South America and then around the world on the HMS *Beagle*. Suggesting that these were formative experiences for Darwin would be a ridiculous understatement. The voyage was life changing, career defining, and it gave him all the materials he would need to formulate his most lasting postulates about "descent with modification." On his journey around the world, Darwin collected and shipped back to London 5,436 animal skins, bones, and carcasses. These were the raw materials that, with the help of others, he would examine in the two decades leading up to the release of *On the Origin of Species* in 1859.

His contemporary and co-theorist of evolution, Alfred Russel Wallace, also spent time exploring South America's natural history. Yet it was Wallace's eight years in the Malay Archipelago that produced his scientific breakthroughs. His experiences and research on these islands, where clear patterns of speciation were evident to the discerning eye, brought him to conclusions very similar to Darwin's regarding natural selection. The two scientists broke new scientific ground by venturing far from their home ground.

After his epic adventure, Darwin turned his talents as a naturalist toward the English countryside, the earthworms in his garden, and the behaviors of dogs and beetles. But the precedent was set. If one wanted to advance scientific understandings of animal behavior, win the respect of one's peers, and comprehend how "from so simple a beginning endless forms most beautiful and most wonderful have been, and are being, evolved," then one journeyed far from one's backyard.

Stay where you are, especially if you work in an urban area, and you ironically run the risk of becoming an exile. This may explain why Seth was scolded by his own advisers during his doctoral exams. Out of the gate, he tells me, this was the first question he fielded: "Why should we waste our time studying weedy animals in cities when that time and money could be better spent working in real nature?"

Real nature, from this perspective, doesn't have anything to do with urban areas. Cities are biological sacrifice zones, full of common species at best, invasive nuisance species at worst. The combative query let Seth know that legitimizing his work as worthy science would be an uphill battle.

We jump back into the truck, which smells of the rank concoction that Seth and his team use to lure roving animals within range of the cameras. He glances at the jar holding the scent attractant and smiles ruefully, "No matter how many times I clean the interior of this truck, the smell stays." I compliment him on his brewmastery. We roll down the windows in unison.

Over the course of the day, Seth and I spotted a coyote trotting across a golf course near Columbus Park; visited a cemetery to pay our respects to the living beings who make use of the habitat it provides; and, at a city park, we played the roles of an ecological Moses and Aaron, parting a sea of Canada geese on our way to a pondside camera.

The time gaps between collecting memory cards affords a lot of time for conversation. This frequently leads to non-ecology-related topics of discussion. Seth is curious about my academic background in religious studies. I hesitate for a beat. Outside a classroom environment, people often conflate the *study* of religion with the *practice* of religion. I lean toward an animistic view of the world, but I walked away from institutional religion a while back. To avoid confusion, when someone casually asks what I got my degree in, I often respond with environmental studies. Close enough. And I did teach environmental studies for two years.

Seth expresses genuine interest, though, and we've got time for more than a hurried elevator pitch. "You study what nonhuman animals are doing. I suppose I've always been curious about what people are doing," I say. "If possible, I'd also like to understand why they're doing it. That's one reason I studied religion, to try to comprehend the mysteries of what motivates people." As Seth gives an understanding nod, it occurs to me that maybe I'm an exile, too. I got my degrees in the humanities. But I have always been interested in the more-than-human. Who do we think of as part of our moral community? Which nonhuman species do these communities include or exclude? I've come to appreciate that answers to these kinds of questions are shaped by our stories about how we got here, what our purpose is, and where we think we're going.

"Our greatest cultural myths were built to provide answers, or at least provisional ones, about our place in this world as human beings," I venture to Seth. I look out the open window at the parking lot of a Jewel-Osco store. "These stories impact how we treat the land. That's one reason I'm interested in your work. Ecosystems and ethical systems inevitably overlap."

Seth takes this in. "Yeah, I can see that," he responds. As an urban ecologist, Seth must account for human behavior in his everyday work. "People come up to me all the time while I'm checking camera traps, wondering what I'm doing. When they find out I study urban wildlife, they almost always have an animal story of their own to share." These stories run the gamut, from concern and care to annoyance and disgust. He does his best to listen to these tales, and then shed some light on why animals are in the city, lessen people's anxiety if he can, even share in the wonder and excitement of others when that's possible. Seth's work involves gathering data, like we are doing today, but his goal is to make the city a better place for all kinds of species, humans included.

As he segues into a story of his own about a crazed raccoon and a pizza box, I begin to feel a deeper kinship with him and his work. In some ways, we are coming at the same goal from different angles, both of us searching for ways to promote living alongside other animals with care and respect—him with a set of data points about urban wildlife, me with a set of stories about them.

We are interrupted by a brief stop at a golf course to check a camera near one of the greens. Golf courses provide great opportunities for urban wildlife: open space, vegetation for food and shelter, water sources, and infrequent human presence. Compared to other landscapes in urban areas, golf courses are often treated as though they are sacred spaces. I wonder if the animals think of them that way.

When we return to the truck, our conversation goes a little deeper down the rabbit hole, touching on nature-venerating religions such as pantheism, animism, and neo-paganism. Seth has a bit of an insider's view of this topic. He had a Wiccan girlfriend in college. Solstices and equinoxes were like Valentine's Day and Earth Day rolled into one.

I begin to wonder why I never dated a Wiccan, but not wanting to lose the thread of our conversation, I steer back toward the topic of how science and religion might overlap. "I think a lot of scientists are more religious than they publicly claim," I say, "even if they aren't theists." I mention the renowned biologist E. O. Wilson, who mused, "If a miracle is a phenomenon we cannot understand, then all species are something of a miracle." Then Seth and I talk about his own field of conservation biology, with luminaries like Michael Soulé, who frankly admits that his work is "infused with morality," and Reed Noss, who calls the discipline a "normative, value-laden science" driven by the belief that "biodiversity is good and ought to be preserved."[2]

Seth swings the wheel to the right and we turn into a near-empty parking lot. "A lot of scientists might not put it this way," I admit, reaching for my backpack on the floorboards, "but evolution is one of the greatest stories in the modern world. Aldo Leopold referred to the self-organizing complexity of evolutionary processes as an 'odyssey'—an epic journey of the universe."

Seth shuts the engine off, and we walk toward the back corner of the lot. I'm not quite sure how—maybe all the talk about pantheism and odysseys or maybe we were both just nerdy enough to allow it to happen—but our conversation turns to a discussion of George R. R. Martin's bestselling Song of Ice and Fire series (the source material for the HBO program *Game of Thrones*). Seth's usual poker face transforms, replaced by a broad smile. He lights up while offering a theory about the parentage of Jon Snow, a key character in the books. He makes a strong case for a Targaryen father (a lineage associated with fire-breathing dragons) and a Stark mother (people of the icy northlands, where always "winter is coming"). We carry on about the finer details of the theory, impressed and surprised by the depth of our mutual fandom. There's a lot of time to kill when you're checking camera traps.

We arrive at one of the most unusual sites we visit this day, a sliver of woods between a seemingly abandoned warehouse (All-Foam Industries Inc.) and a narrow channelized creek. A few steps off of the pavement, we brush aside the branches of a shrub thicket and emerge in a gap. "The TV graveyard," Seth announces, sweeping his arm to his right like a museum docent. Strewn over the ground is the busted plastic and metal shrapnel of several old televisions, most of them face down as if ashamed of their demise. It's a weird scene. The tube screens soldered to gray casings suggest a mid-1980s date range for the dump, a Reagan-era midden for some future archaeologist to scratch her head over.

A train with blue boxcars screeches by on the opposite side of the creek. Tree swallows skim across the water. Seth follows my gaze. "We actually get a lot of mink photographs from this site," he says, throwing a thumb back toward the camera.

I raise my eyebrows. "Really?"

"The camera got stolen a while back, but I replaced it anyway because the site is just too cool." A flicker of excitement passes across his features, and I understand more fully why he goes to the trouble of doing all this data collection in odd pockets of the city. It *is* cool.

When Seth walks off to check the replacement camera, I pay my respects to the TV graveyard, a site that holds rare urban animals like mink, their pliant bodies woven through living currents of water near the shredded circuits of defunct televisions. The city receives, never quite absorbing, the remains of our technological profligacy— story boxes, decaying into soil while a deeper story persists in the nearby creek.

The train clanks by, and my mind again flashes to Thoreau, who watched a train cut by Walden Pond on its way to Concord, under-scoring the rapidity with which the nineteenth-century landscape was changing. After discussing how much labor goes into building a railroad that provides questionable gains, Thoreau concludes, "We do not ride on the railroad; it rides on us."[3] He witnessed the birth of the modern American city in his lifetime. He watched as farm labor shifted to factories, as intercontinental transportation by steamboat and railroad became a reality, and as the speed of interpersonal *communication* increased without building healthy *communities*. "Our inventions are wont to be pretty toys, which distract our attention from serious things," he writes in *Walden*. "They are but improved means to an unimproved end.... We are in great haste to construct a magnetic telegraph from Maine to Texas; but Maine and Texas, it may be, have nothing important to communicate.... As if the main object were to talk fast and not to talk sensibly."[4] I look down at my iPhone. No new messages. I look at the creek.

What stories do we tell that prevent us from seeing what is right in front of us? With each memory card gathered, each image cataloged, Seth is gathering the materials with which to tell a new story about the city. I have a hunch that people such as his PhD adviser will eventually come around and that Seth's sense of exile won't last much longer. I can see the outlines of this new story—one that reveals the degree of entanglement between our urban lives and the lives of other animals in this odyssey of evolution.

The animals of the city go about their business, mostly unnoticed. As do Seth and I on this day. It may not be Darwin's *The Voyage of the Beagle* or Wallace's *The Malay Archipelago*, not the glories of discovery associated with finding species new to science in exotic tropical locales. But Seth and his colleagues at UWI are piecing together an

ecological puzzle, research that will illuminate how we might achieve a more thoughtful coexistence in the city. Another generation of urban field ecologists is coming up behind him, preparing to explore these almost-forgotten wilds. This work, at present, may not confer the bragging rights of field studies conducted in the Galapagos, but a hungry urban ecologist can always count on Tong's Tiki Hut.

De los pajaritos del monte

I celebrate the city and its possibilities. The city provides habitat for a polycultural, polyrhythmic, polyspecies gathering, a place for a vibrant exchange of ideas and ingenuity. Something is always moving, always shaking.

But at times I sense I'm passing through without touching the ground, like a wisp of lakeshore fog. For all its amenities and opportunities, an urban world is a hypermobile world; its currency is speed and convenience. Tendrils of pavement stretch into every nook, promising time savings, even while pedestrians pass idling cars at rush hour.

The city's speed resonates with the transience of my own life. I've been a rambler for a long while, pausing at academic way stations in California, New Jersey, Vermont, Florida, and Texas, crisscrossing the country, never quite settling, accumulating degrees I assumed would create the foundations for a stable life. My geographical promiscuity guided me to many fine places but did not help me grow roots or nurture a strong sense of belonging to a particular place. Then came Chicago and the fresh challenge of trying to dig my heels into urban concrete.

I ponder the disconnection between foot and ground as I ride atop and over the city on the Metra train. On bleaker days, a historical time lapse unfurls in my mind: train after train fed into the blast furnace of the city. The trains become more streamlined as time proceeds; the steel scrawl of tracks multiplies across the land; the people queue at the stations, checking their wristwatches. On the highways, the cars join the push toward the city in single file, bumper to bumper, eager to keep pace.

Chicago has many natural graces, but it also siphons human labor and capital to feed its restless engine. Many writers haven't been able to resist christening the city with monikers to match its expansive aspirations. These, famously, from poet Carl Sandburg: "Hog Butcher for the World, Tool Maker, Stacker of Wheat, Player with Railroads and the Nation's Freight Handler; Stormy, husky, brawling, City of the Big Shoulders." The cocksure might of an industrious city is a myth not easily composted.

I see this clearly some mornings when cooling mist enshrouds the Metra, coughing breathy air over blinkered faces. I know this as I detrain and walk across a bridge spanning the Chicago River on my way to work. If I were to turn left instead of right, I would bump into a building that looms over the water. The top floor of the building displays the company's name in large block letters so there's no doubting who owns this particular piece of property: General Growth.[1] I have not checked the details on what such growth entails. So unquestionable, so good is growth, a company can simply forgo any creative sloganeering and call itself General Growth. If I stare too long at the sign, a knot forms in my stomach about what it generally takes to create general growth. This kind of traffic flow has casualties.

I know the feeling of a train striking a person. A bump. Then deceleration. A clipped squeal of brakes, metal on metal. A stench of burnt toast. Until we stop. At no stop.

The conductor updates us: an "incident" has occurred with a pedestrian. The police arrive to check the scene, as though we all have roles to play. A colder automated voice follows, instructing the passengers, "Please be patient." Hollow air inside the train. We read newspapers, shuffle pages, check iPhones, *tap-tap* screens. A woman in a handsome black blouse glances up at me from her Kindle and shrugs her shoulders. What else can be done? Earbuds remain in their holes. No one speaks into the void.

The automated voice again: "A pedestrian incident has occurred. The train will be delayed. Thank you for riding Metra today."

A lifeline and a commuter line intersect and diverge, carrying all home. Spirits enclosed and still operational; spirits released and gone, gone, gone.

"A pedestrian incident has occurred. The train will be delayed . . ."

Someone had to program that voice, anticipating such incidents. The train huffs, the wheels roll on, the schedule resumes. The blast furnace must be fed. General growth must continue.

"Thank you for riding Metra today."

. . .

What if to live on land, one had to prove one could live *with* land? Not buy it, deal it, trade it, or steal it, but respond to it, show one loved it, become worthy of it? Boxes and grids offer the advantages of efficiency, chances at urban sustainability even, but to relate to a greater-than-human world, to be obligated to particular places, something more, something messier, is required. I am searching underneath the pavement, digging for an obligation to the land's well-being, for richer, for poorer, till death—not the market—do us part.

When I think of land attachment and commitment, a friend whose writing I deeply admire comes to mind. Aaron Abeyta hails from a small southern Colorado town named Antonito, with a population of just under eight hundred people, mostly Chicano. His family descends from a long lineage of shepherds. They are still herding sheep. Although he's now a mayor, my friend continues to take the family flock to graze pastel-colored prairie valleys every spring and autumn, and wild, sandstone-studded, high country pastures in summer.

I envy his knowledge of this land. His family's novenas and curses at the weather, wishes for less snow or more rain while staring up from underneath the brim of a cowboy hat, are experiences sunk into his marrow, common and familiar to him as the warm breath of the sheep or the miniature icicles that dangle from the wool under their chins in winter. His stories focus on the quirks and characters of his fellow Antonitans. He describes his writings as echoes that return from "a voice of hope shouted into a canyon of despair." I never have read someone who dignifies brokenness with such care, who is unflinching in his embrace of those whom others routinely dismiss as unworthy. "The place where I live," Aaron says, "is a place where you know everyone who dies. And you *can* die of a broken heart."

I cannot think of my friend without thinking about the hard and beautiful place where he lives. The mountains there have moods, dark and brooding as shadows of cloud fallen to ground. Intersections and isolations define his place. The migration of peoples and their distance from other Spanish speakers laid down footpaths in the language still retained in collective memory, colloquial phrases, and words heard in his region alone.

A cliff side may appear uniform from a distance, but closer inspection reveals gleaming quartz, mica, and gypsum shards embedded in a mother layer of ocher sandstone. The dialect of Antonito is similar, revealing glittering irregularities within the Spanish spoken there. Many of the older terms survive in the face of "standard" Spanish. Take the Nahuatl word for hat, *cachucha*, which persists in Antonito instead of the more widely used Spanish term *gorro*. Or *bolillo*, a noun from Old Mexico Spanish, which describes a pale French-style Mexican breakfast roll. *Bolillo* also doubles as a word for white people, I suspect not intended as a term of flattery.

To me, a *bolillo*, the most interesting words of this region are the ones that people had to invent themselves to come to terms with their new landscape. Local names bubbled up from observations and interactions with the distinctiveness of place: *cuipas* (shavings of juniper bark), *palomita* (butterfly, moth), *desorillo* (rough pine slab), *empelillado* (describing a horse or donkey that is weak from eating tender grass), or the gem *pelilludo* (describing a horse or donkey with only one testicle).

Even illegitimate children are accounted for by place, the little ones known as *de los pajaritos del monte* ("[gift] of the little birds of the woods"). The land offers language and parentage, birthright and nurturance in the face of the unknown.[2]

In a beautiful fictionalized series of stories about his hometown, Aaron compares the five distinct language sources of his region—archaic Spanish, Nahuatl, Rio Grande Indian, Old Mexico Spanish, and regional New Mexico and southern Colorado Spanish—to the ingredients that create tortillas. He writes:

> There are five ingredients for tortillas. Too often the people think of themselves as only one ingredient. The truth is that the people of Santa Rita do not like to be called Mexican.
> "We don't speak Mexican Spanish."
> "We don't even speak Spanish Spanish."
> "There is no one in the world like us."
> Of all the words, only those last ones are probably true.

He continues by detailing the necessity of each ingredient for tortillas, concluding with the basic gift of water: "Watch as our abuelita, the part of our being that respects the earth, slowly adds water to her mixture, watch as the water makes the table and bolio [rolling pin] necessary, watch as the tortilla spreads itself over the wooden table like water running into a field. . . . Watch the water. It makes us possible. It makes this place and people rise together and not be afraid."[3]

Aaron's commitment to his place is not something he reasoned his way into. He did not deliberate about environmental ethics in a college classroom to learn responsibility or reciprocity. He is bound to these lands as much as he is bound to himself and his family. It colors his thoughts, his skin, his dreams. The waters run through his veins as they run down the mountains into the *acequias* below.

His life has not been an easy one. It includes a wealth of culture and an angry bruise of white prejudice. At a writers' gathering that began to lean toward abstract and philosophical ramblings, I remember the flash of defiance in his face when he said, "There's a lot of words being spoken right now. My father taught me not to trust people who talk too much. People who talk too much are always trying to cheat you out of something." Better to let the words rise slowly from the land. Put an ear to the ground. Listen . . . Listen.

The reason I am thinking about my friend is because of his rootedness. He knows what home is. Even when Aaron is away from Antonito, he is never without it. It is clear to him, and even to me as an outsider, that his people have grown up not just on the land but *with* the land, drawing its sustenance into their vocabulary, their grammar, their speech patterns, and their bodies, which have been sculpted by its forces. Land and people, in a physical dialogue, at a horse's walking pace, a mutual shaping. The arroyos are mirrored in the deep grooves on older men's and women's faces; the ochers and reds of southern Colorado recorded in the palms of their hands and soles of their feet; the wide expanse of a clear winter sky inscribed across their brows.

Aaron grew up with an extended genealogy of names carved into the family tree of place. His birth was not that of a single baby, born in a rural county, recorded by a hospital administrator at eight pounds, ten ounces. An entire fluvial delta fed his coming into the world. He honors those memories by writing, by keeping the water flowing.

Was there ever a moment when he thought he would live elsewhere? Even were he to move away from Antonito, I have little doubt the waters would sing to him, bring him back to the cottonwoods and sky, granite and sandstone, retrieve and absorb him like the piñon smoke of his family's collected prayers.

■ ■ ■

I am thinking about Aaron because I am uncertain about my place, my home. I suspect that he is the exception to the modern rule, that my experience is more consistent with the majority of people who are now of the city.

Or of the country. The rapid pace of change extends its reach. I recently was taken aback when a rural rancher in Colorado complained about the transience of his community. Between the tourists and the struggling ranches and the second-home owners, he believed that the hard-earned relationships between people and land—an expression of beauty that can't be reproduced by scenic postcards—were in peril.

This rancher's concerns were a reminder to me that the pace may be quicker in the city but that no place stays the same. My friend Aaron's home and language are accretions of sediments from various cultures. His memories depend on careful cultivation. His writings reinscribe a people always in danger of being forgotten.

These words I write are my own series of threads to lash myself in place. I don't know if they possess strength enough to hold against the fraying power of the city. I stretch my imagination under the buildings of Chicago, look for ways to peel the paint, scratch at what lies behind the clocks and billboards. I turn toward the animals and plants—to see what they have to say about this particular mingling of wind, water, sunlight, and glaciated ground. I am searching for another language, a music that was here before there were words and will be here still when we have moved along with ours.

I believe a piece of this more-than-human language recently arrived on four legs, a connection between cityscapes, carried across time and into Chicago's present. *Coyote.* Sniff at the word a bit, paw around the edges, and you'll find a small wolf held in high esteem by the Aztecs of central Mexico. In their language, Nahuatl, this indigenous canine was known as *coyotl*.

Something of the extent of the Aztec reverence for the *coyotl* can be surmised not only from codices, carvings, and place-names but also from divine incarnations—such as Nezahualcoyotl, a god associated with music and poetry. When European and American explorers first encountered coyotes, they frequently referred to this more diminutive cousin of the gray wolf as a "prairie wolf." By the mid-nineteenth century, however, the Nahuatl term *coyotl*, slightly Hispanicized to *coyote*, had become the name of choice across the country. Coyotl, who lived near the Aztec capital of Tenochtitlan, the largest city in pre-Columbian America, has since spread his slightly amended name to every city in the United States.[4] As with the dialect in Antonito, Colorado, a Nahuatl word has managed to slip into the story of the Chicago landscape.

Coyotl's descendants now yip into the Chicago air, inhabiting another city of changes, singing about their mastery of adapting to all circumstances. They inhabit this city as they have ever since humans began to build cities on this continent. Coyotl has seen such things before.

The question is not whether coyotes will be with us as the future unfolds. They've proved their skills at long-term inhabitance. For me, the question is whether we will be with them. Can we find our way into belonging?

City or country, perhaps we are all *de los pajaritos del monte*, little birds looking for home, seeking to return the gift we've been given with proper gratitude. Put an ear to the ground. Listen for the music passing between the screeches of the trains. Listen . . . Listen.

II

Anima

Coyote Calls a Council

After a full afternoon of gathering sweets, Bumblebee was taking a much-needed rest on top of a hollow oak log. A small gust of wind picked its way through the trees, wafting a distinct odor down the embankment to her sensitive antennae. Not a moment later, in a panicked huff of motion, Coyote came crashing through the underbrush. He leapt over the log where Bumblebee sat and slumped into a heap of fur, his breathy pants mingled with the distant sound of cars. "You look in sore shape," Bumblebee ventured, eyeing Coyote as he licked a bruise on his leg. "So dangerous," Coyote muttered. "The solid soil carries the metal boxes too fast. One nearly ran away with my leg, and certainly would have if not for my dancing skills."

"Ah," Bumblebee sighed, "I know these dangers. The gleaming boxes seem so far away and then so close all at once. Too late and you lose a wing . . . or worse."

"Something should be done," declared Coyote.

"Something should be done," echoed Bumblebee.

Coyote decided to call a council of animals. Many came to attend. Beaver and Raccoon, Heron and Swallow, Opossum and Skunk,

Bumblebee and Butterfly, Owl and Robin, Catfish and Peregrine, Spider and Wasp, and many others besides. Even Pigeon and Squirrel showed up, casting uneasy glances in Peregrine's and Owl's direction.

The moon was high in the sky when Coyote began to speak. "Brothers and sisters," he said, "thank you for coming. All of you know the benefits of living here or you would not stay. There is good food, plenty of it for the smart and resourceful. There are places to sleep, homes for our children. Water for drinking that never runs out. But . . ." Coyote paused for dramatic effect, casting his yellow eyes upon the gathering. "We must talk about the Humans."

Grumbles, a flapping of wings, and murmurs of assent rose from those gathered. Encouraged, Coyote continued, "Few of the humans seek to find us. Few of them pay attention to us at all. They move as though we are not here. I do not need to work hard to be invisible. They look around and over me, down at small squares in their hands. They talk into the air. Walk without stopping. And, at first, I believed this was good. The best thing for all of us."

Again, the animals murmured approval. "But," Coyote declared, "their metal boxes are deadly. When they put on wheels, they run without caution across the solid soil. Without trails of our own, we cannot avoid them."

Here, the winged nations clucked and squawked. The six- and eight-legged nations rubbed their many legs together. The finned nations flapped their tails. The four-leggeds stamped their feet. Coyote nodded, letting the commotion reach a crescendo, and then barked, "Who will volunteer to teach the humans how to listen?"

All at once, a great net of silence swept across the group. The winged nations stopped clucking and squawking. The six- and eight-legged nations stilled their many legs. The finned nations' tails ceased moving. The four-leggeds were mute.

Coyote burned his gaze into each of the creatures at the council. Many looked away, shamed. Some stared into his eyes, defiant. At first singly, then in pairs, then in large groups, the animals left the council, slipping back into the underbrush, wading into the waters, creeping into the ground, and flapping away on silent feathers.

Coyote stood alone with an empty head. He reached into his pocket and wrapped his paw around his dice. Clicking them against his nails, he pondered the night sky. When he had made up his mind, he pulled the dice out and tossed them high in the air, so high they were lost for a moment in the moonlight. Shouting a single yip into the night, he leapt, caught the dice between his teeth, and ran toward the dawn. Heron, puzzled, watched from the riverbank as he bounded away.

An Etiquette of Sound

This is not a story about the discovery of a new species of frog in Papua New Guinea. To bring back this information, I neither cleaved wine-dark seas nor sailed capes of hope. I did not climb Kilimanjaro or run class-five rapids in a Tibetan canyon. No crampons were used, ice axes deployed, or emergency flares burned. Supplemental oxygen was not required.

I saw a robin for the first time with my ears.

Shake your head now and ask, *Who is this man of high adventure? Robins are everywhere*. Precisely. And I heard one for the first time.

This spring, at our local elementary school, a barricade was erected. Nine caution cones sprang up overnight like mushrooms, creating a fairy ring around a potted tree just outside the entrance to the school. What danger was afoot? The first thought that occurred to me while unloading my son and his backpack into a single-file line was that a broken bottle needed cleanup. The second thought was that a saintly teacher had taken mercy on the tender sapling, sensitive to its long sufferings at the abuses of little hands and feet. But what I discovered was more surprising. A pair of robins had chosen this unlikely spot to deliver and raise their young.

No ecological justification could be claimed for setting up a barrier. The city would not suffer were one robin's nest to fail, nor would the robin population (among the most abundant year-round residents in town) as a whole feel the impact. Other reasons must have been at play.

A piece of paper, Scotch-taped to one of the cones, carried this message:

> Please do not
> disturb.
> Momma robin
> has laid some
> eggs in her nest.
> Shhh . . . quiet please.

This may be the best example of a "teaching moment" that I've ever witnessed—more important than any follow-the-lines printing or counting by tens that my son learned in his kindergarten classroom. Though I understand well the absolute impossibility of quiet children—please or no please—at the front entrance of a school, this message communicated a simple but profound lesson: other creatures are worthy of our quiet respect.

Viewing a nest at eye level is a treat. No sore neck. No binoculars. But, for me, the nest held significantly more than the three eggs it cradled because its appearance coincided with a book I'd recently been reading. *What the Robin Knows*, by Jon Young, makes an arresting assertion that robins know something, and that *something* is at least worth enough to have a book written about it.

The bulk of *What the Robin Knows* is dedicated to "deep bird language," the five vocalizations that are typical of songbirds (excluding corvids' anomalous behaviors). These include the four "baseline" vocalizations (singing, companion calls, territorial aggression, and

adolescent begging) as well as the varied alarm calls that usually signal the approach of a predator. In a fascinating twist, alarm calls can
be used to the advantage of a bird like a Steller's jay—a corvid—who
imitates predator vocalizations and mimics the alarm calls of other
birds. This turns out to be a very useful ruse for diminishing competition around a bird feeder.

The detailed explanations of what these five vocalizations mean,
and the behavior that frequently accompanies them, provides valuable perspective on the power of birds' observational skills. Their survival depends on it. An anthropologist might describe this as Young
does, as a "culture of vigilance" that exists on the streets, in the woods,
and throughout any given meadow. Every place is "full of little eyes
and ears" that obey the unwritten laws of energy conservation.[1]

Take companion calls, for example. Young provides an English
"translation" for the domestic repartee that often occurs between cardinal pairs in the form of a series of *chips!*:

> *Are you there?*
> *Yes, are YOU there?*
> *Yeah, I'm here!*
> *Don't do that to me!*
> *Sorry.*
> *Okay.*
> *Everything all right?*
> *Fine.*
> *Okay.*
> *Dear?*
> *Yes, honey?*
> *You still there?*
> *Yes, dear.*
> *Chip.*
> *Chip . . . Chip . . . Chip.*[2]

Through a series of brief, detective-like anecdotes, Young details
how birds not only are aware of their surroundings but also are aware

of the *particulars* in their surroundings, including types of predators, their directionality, and even their intentions.

Here's the rub: you are one of those potential predators, and you will be treated accordingly. But you need not be perceived as such. This is the point at which the book becomes simultaneously very practical and very philosophical. By developing the skill of "diffuse awareness," "wide-angle deer hearing," or what Young most often calls "jungle etiquette," one can begin to recognize patterns (baseline behaviors) and when they are broken (alarms and different types of alarms). If you want to avoid alarming birds, and therefore increase your opportunities to encounter wildlife, you'd best learn how not to disturb the vocal gatekeepers. This means expanding awareness, learning to recognize patterns of avian language, and acquiring a posture of care; Young sums it up nicely in a phrase: "Softening your presence in the world."[3]

. . .

Robins are common where I live. They swarm the open grounds of my neighborhood's soccer field. They pluck recalcitrant worms from the local playground soil. Their tuneful whistling ebbs and flows between the metallic braking of the commuter train. The robins are common; and therefore, they are uncommonly great teachers, available to all who decide to tune in to their communicative presence.

During the nesting of the school robins, my son and I often visited the school playground after hours. We checked on the nest together, always careful to maintain our sanctioned cone-based distance, and we kept tabs on the progress of the young mouths as they expanded to fill out their woven-twig abode. On one of these trips, we happened to ride our bikes onto school grounds with three other migrating boys that we somehow picked up along the way. That's when I heard my first robin.

I've heard many robins before, of course. In the back of my mind I vaguely apprehended that *chips* and *tuts* and *zeee-bits* had some communicative purpose. But this chatter was indistinct "bird song" to me. As with all background chatter, these noises blended into the larger soundscape of a visually oriented life. Meanwhile, I attended to pressing matters, such as checking for cars at the crosswalk and keeping my child from inadvertently running face-first into a tree. This time, however, because I'd been cued and prepared for it, I heard something new.

As the boys raced each other across hopscotch-scored blacktop, they whizzed too close to the caution cones. *Zeee-bit!* and up from the nest flew the mother. The kids created a perfect bird plow. The mama robin dashed a straight-angled flight line to the low-hanging branch of a honey locust tree. Now at a safe distance, she assumed the sentinel posture, scolding us in a series of agitated *tut*s and buzzy trills. The boys were oblivious. They moved away from the nest on their own accord, preoccupied with their first of many rounds of tag. The robin gradually unruffled her feathers, her suspicion becoming less emphatic, until she determined that our group was indeed a bunch of bumblers with non-robin-related intentions. She quieted, flew back to the nest, and resumed her mothering.

Now, it's fair to respond: any moron should have known what was happening in that moment. But for me, a healing had occurred: the world was speaking and the language was intelligible.

. . .

The same season that the caution cones appeared, my son and I were walking our bikes to the storage room of our building when we noticed a funny-shaped rock with an eye. Closer examination revealed a robin fledgling, huddled in the corner of the condominium's cement staircase. When we approached, the surprisingly large baby took a few erratic skips, contending with legs and wings that were new to the world. My son, moved by curiosity and the urge to help, tried to corral the bird as we speculated on the location it may have flutter fallen from. Eventually I scooped its body—a couple of ounces of fragile fluff—into the cup of my hand. With one foot on a gas meter and one on a handrail, I lifted the youngling up to an electrical-wire box that showed signs of a nest inside. After vigorous protest, the bird settled on some quarter-inch rubber tubing, seeming content to observe the two of us.

My son pleaded that we take the bird home. I assured him, and possibly myself, that the bird had better odds of reuniting with its mom and dad if left behind. His shoulders slumped forward in a sign of begrudging acceptance. Bikes now stowed away, on our walk to the front door my son kept repeating, to himself as much as to me: "He's so precious. He's so precious. He's so precious."

No matter if we inhabit a city neighborhood or a far-flung corner of the earth, avian teachers are everywhere to be heard. Meanwhile, we are carefully watched. We move through well-maintained non-human territories. We are constantly stomping through the stomping grounds of others. How different would a landscape look if the dominant human etiquette were one of "softening your presence in the world"?

I confess, I've never heard a tree say hello. Never had a river speak my name. But I do think the world is alive, resplendent with many tongues. There is probably no end to how I can improve as a listener, as a neighbor, as a large-footed biped in a world crisscrossed by wings, traversed by scaly musculature, and palpated by sensory apparatuses that exceed my comprehension.

I'll begin with robins. They check on me, monitor my presence. The least I can do is return the gesture and gracefully check in with them.

A Language That Transcends Words

In graduate school, when I was doing research about the conflicting values involved in wolf reintroduction efforts, I made a special effort to visit Kent Weber. Kent is the director of a facility that provides a home for socialized wolves—often those abandoned by owners who thought it would be status stroking to own a large wild animal. Kent has over forty years of experience working with wolves and educating people about their behaviors and ecological importance. I have spoken with a lot of people about wolves, but something Kent said during the course of our conversation burned itself in my memory. After detailing the healing psychological impacts his facility's wolves have on various people, Kent synthesized his thoughts: "The biggest lesson wolves can teach anybody: Shut up and be quiet. Get out of your own tunnel and your world and actually take a moment to look at something else that's going to look at you. . . . Wolves contemplate you. They don't just look at you, they look right through you."

The gaze of another animal can have a profound influence, and it doesn't have to come from the eyes of a wolf. Perhaps because humans are visually oriented creatures—dependent on our color-sensitive, binocular vision for so much of our sensory input—engaging with the eyes of another living being in which we can perceive agency, intentionality, and regard triggers a response in us. The studious gaze of one of the Channel coyotes was what hastened me down the pathway to knowing an animate city.

I can think of many other eyes that have since taken my measure—even the eyes of animals who are in the city for reasons not of their own choosing. I consider a recent visit to the great apes section of Lincoln Park Zoo. The Plexiglas, the bustle of people around me, the noise of children—all contributed to my being a distanced observer. I lacked only a lab coat. I watched the animals appreciatively but as I would a television, noting the fine gray, polished lines of their hands, how they ground grasses between their molars, the way they bickered and played with one another. The gorillas seemed near at hand in terms of physical proximity and yet a world away in terms of any genuine relationship between us. Then an individual gorilla turned to face me, and her eyes met mine. A jolt shivered through my chest. A combination of admiration, grief, and awe shook me by the shoulders. The glass between the two of us seemed to disintegrate. I was no longer in observer mode. The eyes spoke with clarity: you are my kin.

Animals with eyes—their similarity and difference from our own—offer an affirmation of our shared creaturely vulnerability, even when these eyes meet ours with seeming indifference. Lessons in humility can be delivered from critters who do not always, or ever, share our immediate interests. Chickens, for example. As the Chilean poet Pablo Neruda mused, in a line that always brings a smile to my face, "I'm tired of chickens— / we never know what they think, / and they look at us with dry eyes / as though we were unimportant."[1]

Indifference can take on much more dramatic dimensions as well and become more physically and philosophically threatening than a disinterested glance from a chicken. The Australian eco-philosopher Val Plumwood tells of the rare experience—at least for a philosopher—of being death-rolled by a saltwater crocodile in Kakadu National Park. Against long odds, Plumwood survived severe injuries from the attack and later wrote extensively about the event.[2] Plumwood elevates her harrowing experience—beyond mere survivor's tale or adventure pulp—by dwelling on the way in which the crocodile's gaze destabilized her sense of self and called into question the "lack of fit" between a narrative of human mastery and a world of indifference to her life and death. In short, she found her ego story ruptured as she glimpsed a "world that would make no exceptions for me, no matter how smart I was, because like all living things, I was made of meat, was nutritious food for another being."[3]

For Plumwood, the dominant story that most of us live by—that humans are radically different from other forms of life, exceptional and exclusive in our superiority, as well as isolated from larger social and ecological communities—shattered against the gaze of another being whose "beautiful, gold-flecked eyes" looked into hers and perceived her as a food item.[4] I can only gesture toward the more profound ruminations embedded in her nuanced and unsettling account, but it is worth emphasizing that she does not draw a lesson of moral indifference from this eye-to-eye encounter. As she notes, a "crocodile-eye view is the view of an old eye, an appraising and critical eye that potentially judges the quality of human life and finds it wanting." Such a view prompts "re-envisaging ourselves as ecologically embodied beings akin to, rather than superior to, other animals," giving us the task of "situating human life in ecological terms and situating non-human life in ethical terms."[5]

A sense of shared ecological vulnerability may explain why eye contact between humans and nonhuman animals is a frequent photographic and film motif. This motif has been interpreted in alternate ways, from alienation to a social contract of moral engagement. Some filmmakers, such as wildlife documentary director David Attenborough, believe that a glance exchanged between humans and other animals can be an entry point for imaginatively considering the world from their perspective. In line with this assertion, many wildlife advocates trace their feelings of moral awakening to what social ethicist Bron Taylor, in his book *Dark Green Religion*, calls "eye-to-eye epiphanies." Nonverbal communication, expressed through the eyes of another animal, seems to reach a place where words cannot.

One of the more renowned instances of this phenomenon is Aldo Leopold's encounter with the "fierce green fire" in the eyes of a dying wolf, which arguably gave the science of ecology its most memorable and influential narrative to date. Leopold's essay about the experience, "Thinking Like a Mountain," describes a turn in his beliefs about large carnivore eradication, a standard government-sponsored practice during his time.[6] The burning eyes of a dying wolf seared a doubt into Leopold, leading him to later advocate for a fuller appreciation of the intricate dance of predator and prey, and called attention to the rippling impacts that human mismanagement can create on a landscape.

Eyes communicate with a language that transcends words. We may be more or less skilled at reading them, but they seem at a minimum to say: We are related. "The eyes break through the mask—the language of the eyes, impossible to dissemble," wrote the philosopher Emmanuel Levinas. "The eye does not shine; it speaks."[7] What that eye says, and what we choose to do after "spoken to," depends of course on the human participant. Denial is always an option. But if we are open to self-reflection, then the meeting of eyes can precipitate a profound reorientation of our moral priorities, even reconstituting our understandings of what it means to be human.

Leopold provides a paradigmatic example in this respect. After a lifetime of ruminating on his field experiences, he was able to more fully appreciate the encounter with the wolf's green fire and translate the ethical implications of that gaze to a larger audience.[8] As he put it: "There was something new to me in those eyes—something known only to her [the wolf] and to the mountain. I was young then, and full of trigger-itch; I thought that because fewer wolves meant more deer, that no wolves would mean hunters' paradise. But after seeing the green fire die, I sensed that neither the wolf nor the mountain agreed with such a view."[9]

Given the transformative possibilities associated with eye contact, it is perhaps unsurprising that eyes feature prominently in religious iconography. In such depictions, the face of a holy figure may seem to follow the viewer when he or she is not directly in front of the icon. In the context of Hinduism, this iconographic gaze has been elevated to a cherished ritual called darshan, the auspicious practice of direct eye contact with a god through the eyes of an icon.[10] Such engagement established through the eyes invites introspection as well as possibilities of transcendence.

The artistic techniques utilized in iconography likely have a biological basis. The expression that eyes are "windows to the soul" suggests we can see into the essence of another person, but what we may be seeing is our deeper bodily kinship. Evolutionary biologist Neil Shubin puts it this way: "When you look into eyes, forget about romance, creation, and the windows into the soul. With their molecules, genes, and tissues derived from microbes, jellyfish, worms, and flies, you see an entire menagerie."[11] This comprehensive kinship speaks to why eyes hold a power to transfix us, cracking apart our self-focus and opening an opportunity for cultivating a more expansive compassion.

I think this matters in cities, where human presence dominates. If we are to care about the nature of cities, nature cannot remain an abstraction. Nature needs a face. Conservation efforts often tout quantity: acres preserved, membership numbers accrued, policies enacted, milestones reached. Quantities are necessary business, the lingua franca of our time, important for assessing goals, charting progress, filling in spreadsheets. But if a person regards a city as a community of life, I'm willing to bet there are particular and emotionally powerful encounters—not abstract numbers—that come to mind. Maybe it occurs in a community garden plot, staring in wonder at the nimble joints of a praying mantis's forearms. Maybe a swallowtail butterfly interrupts one's thoughts about the next email, drifting softly between buildings on an invisible current of air. Maybe a catfish roiling the surface of an urban stream tugs a childhood remembrance of holding that slippery skin between one's hands. Or, more likely, a set of nonhuman eyes—spider, robin, coyote—notices our gaze and regards us, as we return the gaze, transfixed in mutual fascination.

The context of our encounters with these eyes matters. Meeting animals through a screen or on our plates—the most likely place to meet an animal in the city—shapes how we relate, and allows no opportunity for a mutual gaze. Out and about in the city, other urban boundaries constrain opportunities for encounters: rules govern where to walk, where to drive, where to swim, where to sit, where to work, where to play. Nonhuman animals that don't share our sense of orderliness must find places out of reach—whether up in the air or down in the ground or between the cracks. They must be too small to notice or too quick to catch. They must move away or be ornamental enough to accent the scenery—so long as they don't disturb the flow of traffic. And yet the city is an ecological community, a nature we have a strong hand in shaping but a nature nonetheless. Full of eyes, full of faces. Otters, beavers, and black-crowned night herons in the river. Pigeons, honeybees, and peregrine falcons on the rooftops. Squirrels, rabbits, and coyotes in the neighborhoods. There are many eyes upon us.

These other-than-human eyes can shatter the hallway of mirrors that reflects our species-centered biases. These eyes can lure our moral imaginations toward the needs of our kin who call the city home. Yes, the city is a lifeworld we have dramatically shaped to suit our sense of convenience. Yet nonhuman others who navigate that world with us remind us that we are not the only form of life that depends on the basics: soil, water, shelter, food. When their eyes meet ours, we have an opportunity to see what a city is—and what a city could be—with new eyes.

The Cool Red Eye of Chicago

Ever since I moved to Chicago, I've been pondering the question of whether a single animal best captures the essence of this city. Can an animal incarnate a place? On the one hand, I realize my quest is quixotic. I'm tilting at windmills of my own imagination. Which animal best symbolizes a place is one of the more subjective questions a person could ask, and if one were to take it seriously, the answer is open to a thousand viable candidates.

Setting aside the obvious athletic associations of animals with Chicago—da Bears, da Bulls—which are prioritized, as most team logos are, on the fierceness and not the residency of the animal, what might be some qualifications for an urban icon? What suite of qualities lends an animal this status? A certain presence in the way she carries herself, an unquantifiable mystery that induces awe in the beholder over how such a creature could share the same space with us? A rarity that makes sighting him a special event, a reason to run home and write down the date and place of discovery or breathlessly recount the story to others? A charismatic physiology or coloration or set of behaviors that we find particularly beautiful, that pours fresh fuel on the fires of our imaginations?

These questions push beyond "favorites" toward something more ethereal, to the animals we feel especially drawn to without knowing precisely why. Leopold circled such questions and landed on a term for what he was after, *numenon*.

It is unimportant, unless you derive enjoyment from digging into historical cabinets of curiosities, to know that Leopold borrowed the concept, and the ideas it represented, from the Russian philosopher-mystic Pyotr Ouspensky, who, for his part, took it from the turn-of-the-nineteenth-century philosopher Immanuel Kant.[1] The important thing is that Leopold needed a word to describe a *feeling* that went beyond physical appearances. Sometimes we feel things so deeply we grope to see if there is a crayon that corresponds to the color that sparkles in our mind. Germans seem to have a special talent for distilling concepts that would otherwise demand whole sentences of explanation—*Schadenfreude*, pleasure derived from the misfortune of others, or *Kummerspeck*, which literally means "grief bacon" and refers to the excess weight gained from emotional overeating.

Had Leopold lived to see the creation of the field of conservation biology, he might have opted for *flagship*, *umbrella*, or *focal species* as his metaphor of choice for what he was after. But maybe not. Those terms somehow seem too ordinary to be useful. He was chasing something closer to the marrow, something that joined marrow, passed clean through it, and bound it together.

He was articulating something *spiritual*—the numinous spirit of place. Yet he grounded this mysterious spirit of place in a nonhuman animal. Each landscape has a numenon, he writes in the essay "Chihuahua and Sonora," from his environmental classic *A Sand County Almanac*. He proposes that the blue jay is the numenon of the hickory groves; the whisky jack serves this role in the muskegs; the piñon jay for the juniper foothills. "Ornithological texts do not record these facts," he adds with a wink.

For the North Woods, it is the ruffed grouse. About this numinous being, Leopold contends:

> In terms of conventional physics, the grouse represents only a millionth of either the mass or the energy of an acre. *Yet subtract the grouse and the whole thing is dead. An enormous amount of some kind of motive power has been lost.* . . . A philosopher has called this imponderable essence the *numenon* of material things. It stands in contradiction to *phenomenon*, which is ponderable and predictable, even to the tossings and turnings of the remotest star.[2]

It is as though a landscape gathers all its energy together, concentrating its "imponderable essence" into a single species that represents the land's will and desire to be.

The bulk of Leopold's short essay highlights the numenon of the Sierra Madre in northern Mexico, the thick-billed parrot—a flashy, chatty, communal bird that boldly makes its presence known by raising the dawn and scolding unfamiliar visitors. The thick-billed parrot, for Leopold, best incarnated the other-than-human forces that constituted the landscape's unique presence. The parrot—or the grouse, or the jay—manifests what defines a place *qua* place, something that, if we are receptive to it, draws us into a greater mystery and fastens us together.

Can a city, a landscape defined by human presence, have a species that qualifies as its *numenon*? Can one even ask that question of a largely artifactual habitat? I don't know how Leopold would answer. But I can imagine that if he were here, expansive thinker that he was, he would indulge me. He was, after all, a person who consistently advocated—in the classroom, in the field, and in his writings—for the development of ecological perception. He cautioned that a doctorate wasn't necessary for this mental faculty; in fact, an advanced degree might prove a liability, because "the PhD may become as callous as an undertaker to the mysteries at which he officiates."[3] For the person skilled with ecological perception, however, the intricate connections between plants and animals, through time and across landscapes, are available no matter where she or he is located. Even weeds in a city lot can pry open a portal to interconnected fantasias.

The city, though characterized by the deep imprint of human activity, is not dead matter or an iron lung, breathing by mechanical pumps and pistons. Woven into its living fabric are the lives of legions. I think Leopold would argue a little ecological perception can go a long way, even in the city.

So what animal might best embody the *numenon* of Chicago?

Coyotes? Make no mistake, I identify with coyotes on a personal level. They are the quintessential urban adapters. Any-habitat-adapters might be more accurate. They continue to surprise and trick their way across the country as the comeback kids. But there's a catch. They've been able to do so successfully because we're missing wolves in Illinois and have been for 150 years, the species that would have hampered the coyote's dramatic success or suppressed their out-migration from the West altogether. The numenon of Chicago must be, how to say it, more autochthonous, a longer-term resident.

Peregrine falcons? Another comeback kid, back from the brink. After a precipitous decline due to human poisons, they are carving a triumphant story across the sky, reasserting their wild presence. Peregrines also provide a feel-good symbol of humans awakening to destructive actions and lending a hand in the recovery process. Other attributes—built for speed, terrifying in their power, graceful in their precision, stoic in their demeanor. Peregrines have made the city theirs, as though they orchestrated the construction of skyscrapers, using us as pawns to raise the flattened Midwest to suit their elevated purpose. Chicago is one city among many that peregrines have reclaimed. Their geographical promiscuity, however, doesn't make them approachable. They live out of reach and often out of sight. They embody the spiraling ethereal qualities of the numenon, but although they are *in* the city, they are not quite *of* the city.

If not coyotes, "the ghosts of Chicago," or peregrines, the winged dynamos, then whom? I'd like to make an argument on behalf of what might seem an unlikely candidate. A creature with a name that sounds like the sobriquet of a comic book antihero: the black-crowned night heron.

This is a bird of contrasts and juxtapositions. The species takes its name from the slate-colored cap, or crown, that divides the heron's head into a sideways yin and yang, but the first feature you notice is the pair of ivory feathers projecting like white contrails from the base of that head. Or the midnight-blue contrast of his back, offset by the downy white of his underbelly. Maybe the pair of corncob-yellow legs that prop up his football-shaped body and hold it still, as though he's resting on a tee.

Eventually, though, the heron's red eye, a ruby supernova that deepens to a black-hole center, is what will pull you in. This red eye fixes you in its gaze, letting you know that you are part of the heron's passing world, not he of yours. Black-crowned will do, it's evocative as species names go, but better would be the red-eyed night heron.

By land, they don't usually allow me close. If I enter their watery domains, I can, with appropriate gentleness, paddle near enough that the cool red marble locks on to me.

We don't have much by way of rivers in the section of the city where I live. We have to make due with a man-made canal, the North Shore Channel, cut in the early twentieth century. Squint your eyes and it feels like an honest-to-goodness creek. At least the herons think so.

In recent years, I've seen them on the Channel with increasing frequency. They aren't nesting there, but the waterway makes a decent hunting ground. They seem to prefer a section of water near a sluice gate that divides the Channel from Lake Michigan. One day, I saw no fewer than eleven of them perched above the waters, some in trees, some on the gate's cement walls—all of them eyeballing the waters below, meditating on fish.

I felt privileged to bear witness. Night herons are a state endangered bird. They once lived on Chicago's far southeast side in some scrappy wetlands, surviving between steel factories and car assembly plants. Then, around 2009, they moved. They moved toward downtown Chicago. They moved to the zoo.

Location, location, location. Walking distance from Lake Michigan and two and a half miles north of the Loop, Lincoln Park Zoo has nine hundred resident species and serves as "Chicago's Living Classroom." It's also a free-to-the-public zoo, which may account for the more than 3.5 million visitors who come each year. In 2008, the zoo constructed a "nature boardwalk," deepening an already-existing pond that lies just outside the entrance, planting native vegetation, and introducing aquatic organisms. There's a walking path that loops around and away from the fourteen-acre area, a popular spot for strolling and snapping pictures and, now, getting splattered with night heron poo.

The night herons—three hundred nesting pairs, with numbers growing larger each year—have selected two locations for their nesting colonies that overhang pedestrian walkways. One is located by the Lincoln Memorial statue just south of the nature boardwalk; the other is situated above the red wolf exhibit in the zoo itself. I've never heard anyone claim that nests or nest locations are night heron specialties. Unlike the beautiful tight weave of a warbler's home or the mud-dappled engineering of a cliff swallow's abode, night heron nests appear to be an afterthought. Their nests prioritize function over form, little more than a jumble of medium-sized sticks jammed into the crook of a tree.

Whatever works. And, apparently, they will build these brush piles wherever they damn well please. Even over heavily trafficked footpaths in the heart of Chicago. Which returns me to the subject of heron poo. In my estimation, any birds that intentionally or unintentionally put us in our place, cause us to take note of their presence, and remind us that we are subject to more-than-human forces by tarnishing our self-importance as well as our button-down shirts have my respect. Let us call this the virtue of night heron shit.

The black-crowned night heron's construction abilities belie the bird's physical elegance and energetic concision. They are meditators, as I said. The body does not often stir, and when it does, it is for the purpose of mindful stalking. One yellow leg, slowly raised, purposefully placed, the heron makes his food expend energy finding him. Then, like a bolt of lightning, he strikes.

Their nests, their bodies, and their behaviors evoke a prehistoric deep time that contrasts with their modern choice of city habitat. They embody the wild forces, bent but unbroken, that pulse through the city. These attributes offer a solid set of reasons for noumenal consideration. But, for me, the eye distinguishes the black-crowned night heron's claim as numenon of Chicago. We need to be eyed with a bit of suspicion. Our approach should be cause for other animals' concern. We have not gently claimed the city; we steamrolled our way into the landscape, cut channels to flush our unwanted pollutions away from us (and toward someone else), tore through prairie to build shopping malls, and threw concrete wherever we pleased.

And yet.

The herons are part of our story, and we theirs, entangled in a city of juxtapositions that rub against one another, like the black and white of a heron's head. A man-made canal that can't be safely swum but has been repurposed by avian and aquatic beings for food, shelter, and safe passage. A zoo in one of the more densely populated portions of the city that hosts wild nesting colonies, incubating the next generation of herons. An endangered bird who finds the city homey. The black-crowned night heron carries the juxtapositions of the landscape in his body, reclaiming the fruits of modern engineering with a premodern disposition. He is the numenon, the will and self-expression of the land, the mysterious essence of this place. The bird bears these entangled histories, and we with him, into an unknown future.

All the while, the cool red marble warily watches. There are cool red eyes watching us all, wondering if these humans will find a way to adapt to this place, to inhabit a city in a way that is enduring. Red eyes waiting. Red eyes watching.

Vulning

Not long ago, I attended the ordination service of a longtime friend. The event formally confirmed—with song and sermon and ritual—the "call" he had received to be a minister. (And was an acknowledgment of the hoops he had to jump through to get there. This being a Presbyterian service, numerous committees and subcommittees and probably sub-subcommittees oversaw the labyrinthine process to ordination.)

After the service, a group converged for a celebratory dinner at a nice restaurant. Light conversation ensued: *Do you have kids? Where do you work?* At some point, I was asked from across the table what exactly I was doing. I mentioned the projects I lead for the organization I work for, the Center for Humans and Nature, including a project called City Creatures.

"Oh"—an eyebrow arched. "What is that exactly?"

"We're looking at encounters in urban areas between human and nonhuman animals," I replied, "and how these connections can foster care for everyday nature where people live and work."

I'm not sure anyone heard the latter part of the sentence. Conversations skidded to a stop. In my mind, I heard a needle scratch across the length of a record and drop with a hollow thud off of the vinyl.

Human and nonhuman animals.

You know this feeling. The official term is *party foul*. Customary banter about where you live and how's your family, and then someone says the word *orgasm*—and, despite not wanting to notice, a deep instinct arrests all other sounds, akin to hearing a branch snap in a quiet forest—and both you and your conversation partner involuntarily pause. Whether out of decorum or simple embarrassment, you may not turn your head toward the disturbance's epicenter but an unspoken acknowledgment passes between you and your temporary conversation partner. Then, you wait for a moment, you know, just to see what happens next.

Why did I opt for this unnecessary term *nonhuman*? Is it unnecessary?

It's important, I think, to step back on occasion to see what kinds of assumptions are buried in everyday language. I am a stickler for accuracy when it comes to words. I inherited this from my parents. They are a peculiar people, the kind who will go to the mat over the pronunciation of *forte* (they pronounce it *fort*, following the "correct" first listing in the dictionary, instead of the more commonly heard *for-tay*). No one I know outside of my parents—and now my uncle, as the disease seems to be spreading among my family members—pronounces forte "correctly" since this pronunciation throws a verbal monkey wrench into conventional well-oiled conversation. To their credit or discredit, my parents would rather sound incorrect than be incorrect. For the most part, I regard this as a benign quirk, an almost charming affectation. There's not much at stake, other than internal pride. But a judgment lingers around the edges of such pronunciation wars: I know what's right and if I seem wrong, that's on you.

I thought about this as I drove home from the dinner—by saying "nonhuman animal" was I only proving that the apple didn't fall far from the tree? Was I ungraciously trying to drive home a point—that everyone should recognize and embrace a shared animality across species lines?

. . .

I've become accustomed to the term *nonhuman animal. Animal*—without the *nonhuman* qualifier—is a label that creates distinctions, usually negative ones. To call someone an "animal," for instance, is an invocation of baseness ("He behaved like an animal!"), or a slur ("You eat like an animal"), or simply a dismissal ("It's *just* an animal").

Nonhuman animal is problematic in its own ways, of course. It feels unsatisfying to describe all other animals by negation (with the *non* preceding the *human*). Perhaps the only place this negation works well is in the entire phrase "human and nonhuman animals." And still pride of place falls on our species because the pairing implies a division between us (one species) and millions of them (other species). Our language seems inadequate to the task of holding animal continuity and otherness in the same word. We need a way to distinguish between human and nonhuman animals without completely distinguishing.

The people around the table that night probably did not think much further about what I said. Likely, my faux pas was forgotten when the dessert sampler arrived. But I did think about it.

I thought back to the ordination service itself and wondered whether anyone had noticed the animals hovering around the edges of the service, scampering through the text and music. Take the biblical reading, for instance, a passage about Isaiah being chosen as God's prophet: "I saw the Lord sitting on a throne. . . . Above it stood seraphim; each one had six wings: with two he covered his face, with two he covered his feet, and with two he flew." Then there were the hymns, such as "Joyful, Joyful, We Adore Thee." In the second verse, the congregation belted out all the "works" that contribute to God's glory: "Field and forest, vale and mountain, flowery meadow, flashing sea, chanting bird and flowing fountain, call us to rejoice in thee."

I pondered how it is almost impossible to describe an invisible god without reference to a living earth and the animals that dwell upon it.

I once encountered an unusual and powerful expression of this phenomenon at an Episcopalian church in Gainesville, Florida. At the front of the sanctuary, above the altar, a wooden carving of a pelican enfolds her chicks with protective wings. I have seen plenty of stained-glass lambs and fishes in churches, even an occasional lion or ox, but I couldn't recall ever seeing a pelican—especially one occupying such highly valued iconic real estate.

A little sleuthing was in order. As far as biblical references go, I found that the pelican sculpture resonates with verses in Psalm 91 in which God offers feathery protection under wing. But the more direct referent is to a common belief in the Middle Ages that pelicans fed their young with blood from their own breasts, just as Jesus sacrificed his own body for his followers. A pelican Christ.

I was surprised to discover that a very specific name describes this act of pelican self-sacrifice: vulning. When I later spoke to a woman who works at the church, she told me that the pelican mother with her chicks was a symbol of the Trinity. In other words, the mothering pelican represented divine interrelation, the encircling sacred.

■ ■ ■

The vulning pelican is a beautiful symbol, pointing toward mysteries beyond our ken. But what of mysteries with our kin? I mean here not animal carvings that *represent* the divine, but the *actual presence* of animals with whom we share this earth.

There's fairly widespread resistance to this kind of kinship thinking. I suspect the resistance is related to what caused the dinner-party guests to balk at the phrase "human and nonhuman animals." It wasn't simply unfamiliar terminology that prompted the discomfort but the suggestion that humans are, indeed, animals.

We tend to see our species as "uniquely unique" and our humanness as something that needs vigilant defense.[1] This frequently results in definitions of humans *in opposition* to other animals, with the gold standard being our cognitive capacities—our abilities to imagine, reason, and forecast possible futures. Despite mounting evidence that our species holds no monopoly on prizes in the cognitive talent show (along with language, tool use, empathy, self-recognition, morality, and other once-defining human excellencies), it still remains all too common to act as though humanness is next to godliness.

There are and have always been alternatives to sorting out what it means to be human. In contrast to René Descartes's famous philosophical postulate "I think, therefore I am," which defined existence according to cognition, the anthropologist Nurit Bird-David argues that a relational epistemology—characteristic of many hunter-gatherer peoples—is based on the supposition "I relate, therefore I am" and "I know as I relate." Particular relationships between people and other animals are thus based on a "mutual responsiveness," which can grow into "mutual responsibility."[2] Another way of saying this: we *become* human through learning how to relate well to others.

We live and become who we are—as individuals, as a species—in relationship.

Engaging the mystery of Otherness while exploring the kinship between species, we can approach the world as though it could speak, because it may *if we listen*, and it does according to many who cultivate this kind of attention.[3]

In Western industrial societies, we might be allowed such feelings of mutual responsiveness as children but are often encouraged to put such childish notions behind us as we grow to adulthood. That's why I'm compelled by the stories of those who have struggled to overcome the Cartesian hangover, those who have been told that their objectivity will be compromised if their sensual experience of the world informs their treatment of other beings.

The ecologist Stephen Trombulak, for example, writes of how his early love of natural history was supplanted over the trajectory of a scientific career, until he realized that "in the process of composing a life, the natural world had become a subject, scarcely more than the raw material from which I crafted my career. Simply put, I had allowed it to become separate from my spirit. . . . I believed that my profession asked me to treat other species as subjects and not as neighbors, and I complied." The turnaround for Trombulak came from his realization that he was "a citizen of the natural world not just a scientist" and that "human-people share the world with deer-people, fish-people, plant-people, and even rock-people."[4] One learns how to behave appropriately by engaging this world of persons as "neighbors." I relate, therefore I am.

. . .

On an early October day, I was fortunate enough to kayak on the Des Plaines River, near McKinley Woods. My paddle trip happened to coincide with the migration of white pelicans through northeastern Illinois. Pelicans in the Midwest? I had a hard time believing my eyes. Yet there they were, bobbing close to my kayak, glorious foot-long, pumpkin-colored bills and all. I floated in silence as a faction of them unfolded their nine-foot-wide wings and hurled their bodies outward, slapping the water with their feet to gain necessary lift, then gliding upward on supportive currents of air.

As I watched, an image flashed in my mind: the wooden sculpture, in which God is depicted as a pelican, feeding a brood of chicks. If a pelican sculpture can shed light on how to better understand our relation to the sacred, actual pelicans allow us to marvel at a world full of nonhuman movements and consciousness. Which is perhaps another way of saying: how to better understand our relation to the sacred. Pelicans are living kin, full of complexity, with a genome and a history related to our own but with a morphology and a way of navigating through space that is vastly different.

This splendor can be had in the here and now, not just the hereafter. And here they are, unmediated—at hand and too far away to touch, familiar and mysteriously other, human and nonhuman.

Pelican vulning turns out to be a bit of folklore, a mistaken apprehension codified in Christian art of the Middle Ages. The legend perhaps arose due to the visible blood vessels on a pelican's throat, or because pelicans macerate fish in their pouches before transferring this food to their young. But in a way, there is a broader truth embedded in pelican iconography. We live on a vulning earth, and as all creatures do, human and nonhuman, we depend on the daily bread it provides.

We need not know what a "relational epistemology" is to care

deeply about how we relate to nonhuman others. Responding with gratitude to a living world, a vulning earth, is open to all comers.

I wonder, would I describe the City Creatures project differently if I had that dinner to do over? Probably not. I don't think this would be due to my own hardheadedness. Hardheadedness is not really my *forte*. I think it has to do with wariness about reducing other species so that we elevate our own; that's a zero-sum game that leads no-where. Animals we are and nothing is *just* an animal. We all deserve more than that.

Pelicans and other creatures may not care if we regard them as neighbors, but what if we actively choose to do so? I relate, therefore I am. That is a communion worth exploring.

The City Bleeds Out
(Reflections on Lake Michigan)

The best way to live
is to be like water. . . .
One who lives in accordance with nature
does not go against the way of things
He moves in harmony with the present moment
always knowing the truth of just what to do
—TAO TE CHING, verse 8

Gulls puncture the wind. On Chicago's edge, across this inland sea, I know Michigan rests somewhere beyond the line where chalky sky rubs against fresh water. The city loosens its grip here, bleeds out. Steel, granite, limestone, sand, shell, bone, tendril, hand—all will eventually surrender, become liquid.

The lake offers a pause. A curve of shoreline where the urban dweller can contemplate a humbling expanse beyond the buildings. When I come to the shore, I am overcome with gratitude that humans, so far as I know, haven't found a cost-effective way to build upon open water. The water says: no further. With that pause comes an opportunity.

The *Tao Te Ching* provides this counsel: be like water, it overcomes in the long run. "Nothing in this world is as soft and yielding as water / Yet for attacking the hard and strong none can triumph so easily / It is weak, yet none can equal it / It is soft, yet none can damage it / It is yielding, yet none can wear it away / Everyone knows that the soft overcomes the hard and the yielding triumphs over the rigid."[1] Ten thousand years from now? The city will crumble into Lake Michigan. It is the way of water. It rises, it falls, seduced and repulsed by a lunar affair, attentive to cyclical rhythms of which we are only dimly aware. People think you must go to the mountains to encounter wildness. There is nothing wilder than water. Even cities bow before water.

Released from the housing bust, construction has resumed again in the city. Downtown, anything close to the Chicago River serves as a hot spot for new buildings. Few distractions slow the momentum of a businessperson on a cell phone during a lunch break. But with the buzz of construction near the river, it's different now; people stop and stare. They congregate in small groups to look at cranes, soak in the sound of jackhammers, or watch the sparks fire from the hand of a welder. There is a dazzle to shining steel, muscled ironworks, and monuments of concentrated carbon. The motives driving the creation of these astonishments are simple. We network, drill down, and leverage for reasons similar to our ancestors': to cut the bite of the wind in halves and put dinner on the table. And so, we have pulled up soil, practiced chemotherapy on the land, squeezed the blood out of glacial stones, and sent the slag down the river to the Gulf, proud of words like *ingenuity, progress, scale, growth*. Other words are liquid whispers: *love, community, respect, care, obligation*. The water swirls on, above and below, brooding over the void.

The temperament of water is felt most directly on the restless edge of the city. I watch the lake's moods swing with the seasons. Storm surges releasing froth-violent waves that sweep cars from Lake Shore Drive. Ice floes that extend the shoreline half a mile, splintering aged wooden piers, carving out massive hunks of land. A quick shift of season and you will find surfaces so summer-smooth you'd think you could walk to Canada on a blue highway.

Water is a fitting symbol for Tao—without water we wither; it is the substance that seeks a bottom only to rise and recirculate in our every joint. We have saltwater in our blood, freshwater in our spine. The human body is about 60 percent water. Nearly 70 percent of the earth's surface is water. We are water tasting water, mystery tasting Mystery. Lake Michigan flows through my veins. It is the stream from the tap. I wash my dishes with it.

A city can take a toll on water. I ponder this when I find a beer bottle washed up on the shore. No message inside, apart from the emptiness of someone who mistook the lake for a dump site. But I also find fragments of sea glass, green and blue and brown shards shattered and pounded out and rubbed smooth as polished stone, the edge of our indifference made to shine like a gem. Water is as hard as it is forgiving.

Deep time awakens by a sea or lakeshore. Grains of sand and discarded shell homes, honed to a twinkling iridescence: one enters the wormhole of eons. The gulls play court jesters at this dance macabre, with thin laughing calls. Their generations have plied the shore long enough to be in on the open secret: all of us eventually go back to the water for polishing—with hopes of iridescence. "The world is nothing but the glory of Tao . . . / Rivers and streams are born of the ocean / All creation is born in the Tao / Just as all water flows back to the ocean / All creation flows back to be Tao."[2] Eons from now, what form will the feet take that pick up our fragments on new shores? Will they resemble feet?

■ ■ ■

Edges and cycles. On January 1, when calendars shuffle and attention shifts to resolutions and hangovers, I've taken up a new habit. I travel, not to other countries. I get out and walk, not always advisable or comfortable in the heart of a Chicago winter. But I take the arbitrary mark of a new year as a fulcrum and use it to get beyond pavement. Winter sausage and a pretzel roll, a daypack and a deserted train platform—a near-never happening—and I'm off, borne northward to a lakeshore refuge. Neighbors are throwing out Christmas trees, wondering quarterly worries about whether the economy will be beneficent in the coming year. At the lakefront, clarity arrives: there are time scales and then there are time scales. The Silurian, the Ordovician, the Triassic, the Quaternary—massive cyclical expansions and contractions of life. Ebb and flow. Dissolution and renewal.

We have to work to bend our minds around such durations. Familiar objects make unfamiliar scales comprehensible. Here's one way: If I throw my arms as wide as I can, try to embrace the wind pouring off the waves, and these outstretched arms represent the span of earth time, 4.5 billion years, then a nail clipper will suffice to cut human habitation off the body's map. Mind blowing, for the only species—so far as we know—with the minds to contemplate such metaphors of measurement. Sometimes I feel as though the water coughed up a grand epoch of mammals as an experiment, to see what we would carry forward into another cycle in the spiral of time.

To Lake Michigan I go to think bigger-than-me thoughts. To see my own limits transgressed. To imagine what water—the creatrix of land-born life—has to say, has to take, has to give. I come to dip my toe in the edge of where human control falters, fails, and falls into water.

On edges, feet crunch upon shattered shells strewn like a pulse before the water's persistent tongues. I dig bare toes into amniotic sand and grasp grace by kneeling, dirty kneed before a rising sun.

. . .

When I was in my twenties, I spent a year in northeastern Oklahoma on the Osage Indian reservation. The Osage people have a name for god that seems an honest assessment to me—Wakontah, one translation of which is *Great Mystery*. The Tao, Nameless Simplicity, numenon, Wakontah—they are all handles for a smooth door. "So deep, so pure, so still / It has been this way forever / You may ask, 'Whose child is it?'—/ but I cannot say / This child was here before the Great Ancestor."[3] The mystery child is incomprehensible but can be felt, touched, heard by the receptive heart and mind. If you take your net out, you will not catch it; you must be still and feel it tickle your ears and tousle your hair.

And seep between your toes. I walk on damp sand, marveling at what the lake returns to land. The spiral shells are the ones that astound me—the precision of the curvature, the love affair of genome and its expression in mineral homes, scrives a tally mark in favor of the Great Mystery. Surely a million other shapes would work, be as functional for survival. I tip my hat to the Mad Hatter of Life, Wakontah, who deemed the spiral a good shape to pitch out into the universe. A rococo touch from the strong hands of water.

■ ■ ■

I come to the shore when I need to breathe. When I need to have something breathe me. When I need a visceral reminder that the city doesn't have the last word. Water is patient.

The gulls are here again, my only visible animal companions. They circle with raucous confidence and wing me to America's western edge, where, half my lifetime ago, I kept company with their brethren whose eyes sidled my way. The open water and the gulls tug me between lives, between seas—the Pacific, Lake Michigan. Lake Michigan, the Pacific. I have a story about water.

Tides and a full moon—I was immediately drawn to her when we met. We were on the coast of Southern California. We were in college. She was self-possessed, had her bearings, knew injustice when she saw it, and her heart was giving. We danced around each other for a few weeks, met for coffee at a bookshop, went to a concert with a group of friends, then made plans for just the two of us.

We sat on the edge of America together one night, the only inhabitants on a shadowed beach, with the wind whipping through her dark brown hair, flying around her shoulders, her high cheekbones. Our feet were buried in sand warmer than the air. We clasped hands. I can see her thumb on top of mine, slowly drawing spirals. Her chin is to the wind. She is strong, mysterious, beautiful. I am falling in love.

Love is like water. We think it is like rock, but it's not. If we commit our life to another, we commit to water. We say vows as though pledging to rock, thinking that it won't move, that it can be built upon without worry. Water fissures this belief and asks, can you change with a person through their changings?

. . .

Can you expect answers from water? Many years later, in a desperate moment, having fled my gods long ago, I cast about for some portent, heaving a plea for guidance into the waves. I held the ocean responsible for bringing my truest love and me together. I needed the big water to keep its promise.

A gull pondered me, a quizzical look in his red-rimmed eyes, wondering if there were crackers in my backpack. I had none. He tacked left, then right, faced me again. I thought he might have a message to deliver from the ocean.

From ancient Greece to Tibet, India to Polynesia, birds are messengers, flying like dreams between our earth-bound existence and the realm of sky. *Augury*: the practice of looking to birds to know the will of the gods, from which comes the word *auspicious*. We take the auspices, grope for indications of benevolence.

Even within cosmologies that feature more anthropomorphic deities, the messenger birds persist—or parts of them do—in the guise of angels. Real birds, however, possess the advantage of being more visible than most angels, and therefore more consistently available for interpretation. And humans have interpreted everything, from scapulae to entrails to boiled chicken feet, on the lookout for hints about a prosperous course of action.

The gods' answers never come in plain talk, so some cultures developed highly specialized and elaborate systems, consulting the flight patterns, calls, and individual behaviors of birds, codifying these in formal charts and sacred texts.[4] In traditional Iban society in Borneo, for example, birds play critical roles in the lives of the people. Particular birds are believed to be gods (sons-in-law of the high god, Singalang Burong), and any activity for the Iban, from feasts to home building to warfare, is best accompanied by some assurance of avian favor.[5] Still, the Iban must interpret the signs correctly; they must strive to understand the language; they must treat the birds with respect and make appropriate offerings. And yet there's debate; in fact, "debating at length the relative merits of differing augural interpretations is an honoured and favorite pastime."[6]

All of us, Iban or not, need assurance from somewhere, a favorable nod in our direction, an indication that we are on the right track. Think of augury as ethology with benefits.

Do you have a message for me? I ventured cautiously to the gull. *I don't need to fight a war or plant a field. I just want to know how to mend what feels as though it is breaking.*

The bird paced between me and the ocean, paused, offered a slight turn of his head. I waited. The ocean murmured. The gull remained silent. *Last chance,* I said, daring the bird.

Soon it became clear I'd get no magic. Gulls have their own business to attend to, their own company to keep. I wanted answers—from the gods, the earth, nature, our animal kin—some backing, some revelation from outside to quell the doubt within.

I instead received a different message: The world might indeed speak, but it doesn't speak to me alone. The gull is full of its own gullness. If that's not enough for me, I'm asking the wrong questions.

So, we looked at each other, the gull and I. We shared the beach and contemplated the briny smell of the wind together. The ocean murmured. I exhaled. *This is enough. Why ask for more when so much is given?*

. . .

Summer in Chicago. I cannot sleep. The light wakes me early. Mourning doves cry outside my window, plaintive *coo-ooo*s that sound like a question.

I get up, ride my bike to the lake. I come to the beach to remember that feelings are older than humans, to have the water inhale and exhale me.

The water is smooth as a shard of sea glass. I sit in the shallows, water above my thighs, rocking gently with the pulse of the lake. Everything appears to be made of light. The reflective gleam of the rippling wind. The minnows that divide and merge in silver streaks around my knees. The rising sun that warms my back as I cry. Cry because I feel like one thing with the lake. The lake holds me. A mourning dove calls.

A shoreline is a place of constant change. Large waves are called breakers. The result of their work is in plain sight. A shoreline is a place where what is broken becomes beautiful. The work of water.

Ideals break. All ideals must. Our closest loves break. The expansion of love requires this. A shoreline is a place where what is broken becomes beautiful. Perhaps a loving relationship consists of equal partners, rubbed smooth and softened, like a pair of tadpoles in a circle; a yin and yang, each having the other buried in its core, incomplete alone.

We circle back—not circle, *spiral*, for we don't touch down in the same place. Each spiral we turn and ask the questions in new ways. We circulate from river to ocean to sky. We break and our edges soften, polished like sea glass. Two shards washed upon the shore, embracing each other. The work of water. I have collected what the edges of water offer: driftwood, sand, sea glass, the feathers and complaints of gulls, and the unremitting work of waves. "The best way to live / is to be like water / For water benefits all things and goes against none of them / It provides for all people / and even cleanses those places / a man is loath to go / In this way it is just like Tao."[7]

A large body of water is a great mystery. We cannot see too far under the surface. We cannot swim too long without a measure of danger. Where there is mystery, humility can flourish. If we think we fully know something or someone, we are likely to stop fully listening.

Part of the gift of water is that it reminds us of what we do not know. Perhaps this is why I come to the shore to listen. The waves su-

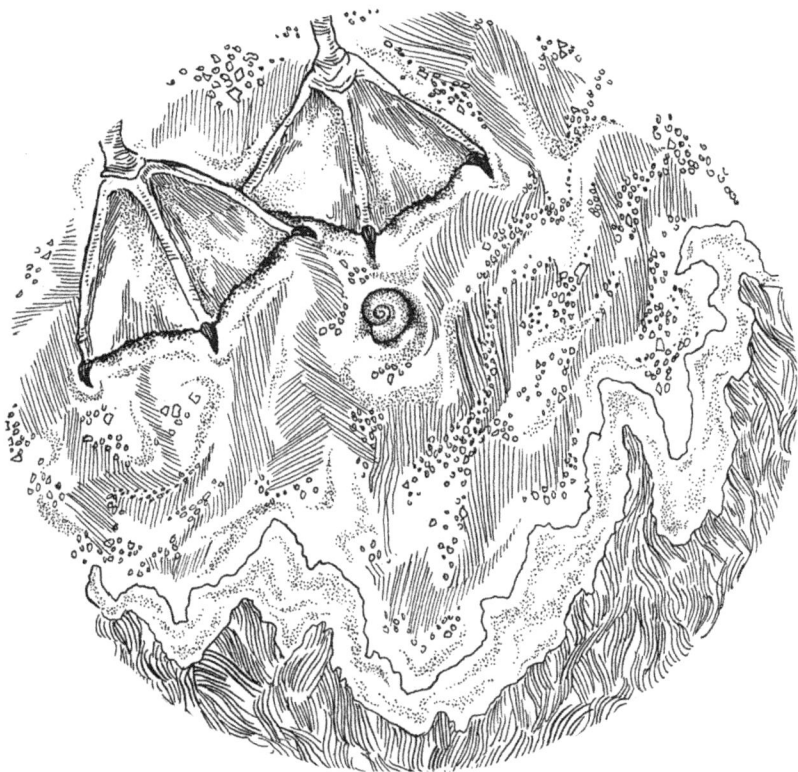

surrate a deeper knowledge, they sweep away my circle of thoughts, and sometimes, I think, I can hear the edges of the mystery that enfolds us.

I watch as the gulls dive and chase one another. On the shore, one throws her head back and squeals at the sky. The only other sound is the gentle lapping of the waves. I dip my finger in the shallow water and trace a spiral in the sand.

Great Blue Meditation

"Emptiness" was neither absence nor a void. Its Chinese character was ku, which also signifies the clear blue firmament, without north or south, future or past, without boundaries or dimension. . . . In this universal or absolute reality, there is no holiness (nor any non-holiness), only the immediacy of sky as-it-is in the present moment, with or without clouds or balloons, kites or fireworks, birds or snow or wind.

—PETER MATTHIESSEN, *Nine-Headed Dragon River*

A footbridge crosses the North Shore Channel. Swallows fly, helter-skelter under and over me, chasing their dinner. Below is a great blue heron. He is still as the cottonwood trunks at his back. Swallows are creatures of tempestuous wind. They outmaneuver and outspeed their quarry. The great blue heron contemplates water for a living.

■ ■ ■

My feet fall asleep in the lotus position. I do not know if this happens to other people.

When we moved to Chicago, I sporadically visited a Zen practice center on Tuesday evenings, "beginner's night." I appreciate that everyone, not just permanent residents and teachers, must wear a robe. I slip my arms through the sleeves, fiddle with the sash around my waist until it holds, walk upstairs. Palms together, bow once to a statue of the Buddha, place my circular cushion (*zafu*) on a rectangular mat, take my seat, twist my legs into a half-lotus position, begin counting exhalations, wait.

Wooden blocks (*han*): *Clack. Clack. Clack.* Handheld bell (*inkin*): *diiiing.* Now breathe, eyes slightly down, facing the wall. Twenty-five minutes of seated meditation (*zazen*). One, two, three, four—*Is my back straight? Am I starting to slouch? Ah, stopped counting.* One, two, three, four, five, six—*I wonder what enlightenment feels like? Is it a completely recognizable moment or something that comes on gradually like the Doppler effect? Why is my stomach burbling? Oh, yeah, nachos. Dang it. Breathe.* One, two—*If nothing else, there is value to just sitting still. I run around too much. All these thoughts bouncing around inside my brain right now, where do they even come from? Am I in a dissociative state? Wait a minute: there is no "I"! But then who is on this cushion? Who is asking a favor of his wife to take bedtime duties while he runs off to sit for an hour? Shoot.* One—*You know, is this something I can even do once a week schedule-wise? Should I just practice at home? I*—*diiiing*—*People are standing. Get up.*

The five other people stand in front of their cushions facing one another, right fist pressed against the left hand's palm. I scramble to do the same. This is when I know for certain that my foot is dead. Tingling pains shoot up my leg. I balance precariously. People move forward and turn ninety degrees to their left. I pirouette, my dominant foot lagging behind like a cinder block latched to my ankle. *Shit.* I stumble, stutter step on the insensate foot, tangled on the hem of my brown robe. I half pitch forward, looking as though I'm trying to curry favor by making an extra bow to the Buddha—for good measure—dangerously close to tipping into a full face-plant. *Miracle.* I do not collapse. I right myself and drag-sway, Igor-like, attempting to close the gap between my body and the enlightened person in front of me.

I did not expect instant insight when beginning *zazen*. But I also did not expect the panicked possibility of falling on my face. If the others notice—how could they not?—no one lets on. Perhaps they are too deep. I am just another passing thought to be discarded. Breathe out.

Five minutes of walking (*kinhin*). People stop at their cushions, face one another, bow, reposition themselves on their seats, twisting feet over thighs. I am one beat behind on every movement. Robes rustle. Knees pop. My foot is finally reattached to my body, circulation resumes. I digest what happened. *Okay, breathe. You're back in a safe space.* One, two, three, four . . .

A bell rings out, the signal for *dokusan*, the time to meet with the teacher. People scramble from their cushions, queuing, then scurry downstairs where the teacher awaits. This night I do not go with them. I will do *dokusan* in the future. This evening, the prospect of asking a question of the teacher feels immodest to me. I have reached my quota for blunders. Safer to stay seated.

Chants close the session. The Zen style of chanting is a monotone hum, words blend into one another following a steady rhythm—one, two, one, two, one, two. The chanting soothes and awakens simultaneously. There is a certain line that stands out against the others, a certain impossible comprehensive ridiculous ambitious aspirational assertion of fact: "All beings without number, I vow to liberate."

It's easy to read that and shrug your shoulders. When you chant it, though, you want to believe it's possible, that you *will* save all beings. It's the crux of Zen to me, the bodhisattva vow: boundless compassion for all life. In forgoing one's own self-interest, or rather, discarding the notion of self altogether, one seeks the liberation of all beings. But it's an absolute mystery what exactly that means. Staying steady on my feet is enough of a challenge.

■ ■ ■

The heron knows. Steady on his feet. He knows the water. Measures it out in feathered breaths. Some say the water raised him up, extended his legs like cattails, and placed a body the color of its various dark

moods on top. Then the water rippled the bones, extending their dendritic reach, for water can riffle where it will—and water wanted to fly.

. . .

The second time I attend beginner's night, I make a decision: I will
go to *dokusan*. I am supposed to have a question for the teacher. I am
nervous. I try to focus on my breath.

I enter the room. The teacher discerns that I don't know the protocol. "Bow toward me." I do this. "Sit," he says, smiling.

I ask if I need to sit in the lotus position. My foot has just stopped
tingling.

"Any way you like," he says. I don't sit in the lotus position. "Do
you have anything on your mind?" he asks, leaning toward me ever
so slightly.

"Non-attachment," I say. He waits. "I don't get it," I say. I pour out
a headful of thoughts: *I can understand not clinging to results, not doing
for the sake of your own gain, not dwelling on the past or future, focusing
on the present. After all, the present is all we have. But not being attached
seems dangerous. Isn't love attachment? Shouldn't we love this world? Isn't
that compassion? Shouldn't we fiercely love our families? Wouldn't we be
less-than if we didn't? Is the ideal of non-attachment a product of Buddhism's monastic context—because one had to forget the world, relationships, leave everything behind to be focused on the liberation of the self?
But shouldn't we respond to need, to pain, to hurt? To care about this world
is attachment, isn't it, and isn't that exactly what we should do?*

He nods slowly. "Do you have children?"

"Yes, a son."

"How old is he?"

"Five."

"When he cries at night in his sleep, what do you do?"

"I go to him."

He nods again.

. . .

Soyen Shaku, who was the Zen Buddhist representative at the World's Parliament of Religions, held in Chicago in 1893, said: "It is the philosopher's business to deal with dry, lifeless, uninteresting generalizations. . . . True Buddhists do not concern themselves with propositions such as these . . . [but] endeavor to reach the bottom of things and there to grasp with their own hands the very life of the universe."[1] D. T. Suzuki, who perhaps did more than any other person to make Buddhism intelligible to those in the West, echoed this thought: "Zen must be seized with bare hands, with no gloves on."[2] It's as close to you as breathing. The gloves of rational thought must come off. The academically inclined mind, one used to poking and prodding, spurred on by constructing refined arguments, must cast aside the idea that the Real can be seized by some formula, that the kernel will be reached by cracking away with logic. It is an ego trick to think one can grasp living truth with words, which build walls around it. Being right is not always right being.

. . .

I am grateful for the indifference of birds, grateful that they go about their business and need not be impressed or catered to, grateful that they eye me suspiciously, then fly on, tracing insouciant loops from roof to roof, from housetop to church-top to trash dumpster, and back again. I am grateful that the world rolls on, tracing insouciant loops through the dark spaces that my eye cannot reach, spinning casually on as stars explode and galaxies without names go about the work of living dying being reborn. The galaxies and the birds take no mind of me, and I am grateful to be a speck of carbon fire blood water that slips along noticed and unnoticed.

. . .

Despite best intentions, it is difficult to commit to *zazen* practice and be a parent. I know there are those for whom these two activities are compatible. I am not one of them. I couldn't muster the discipline to parse my hours, so my tenure at the Zen practice center was a short one.

Had I stayed longer, I would have received a *koan* as part of my practice. A *koan*, a technique developed extensively in Rinzai Zen for breaking through rational thought, is a logically unanswerable mini-puzzle intended to encourage students to grasp the truth without concepts, with bare hands.

A student can work on a *koan* for weeks, months, years, meditating intensely on the question until it becomes a singular focus, until it becomes his or her breath. In *dokusan*, the master questions the student, probing, trying to lift away the layers of conceptual detritus, the desire to reach the answer by reason, to be "correct."

Before modern sensibilities made their way into the zendo, a master could attempt to awaken the student in ways that would now seem extreme. He might listen to the student's "answer" patiently, then strike the pupil across the face. Anything to help his pupil seize the answer, not with the mind but with his whole being. We now rightly balk at such techniques. We've witnessed too many abuses of religious power and people at the mercy of rotted hierarchy. Besides, maybe the master had a bad morning, maybe the zendo ran out of tea.

Still, if we conceive of the slap in the face as a metaphor, then the earth is full of *koans*, pregnant with bursts of potential enlightenment, opportunities to understand ourselves in other selves, dissolving the stranglehold of ego. My favorite Buddhist maxim captures this idea: "Before a person studies Zen, to him mountains are mountains and waters are waters; after he gets insight into the truth of Zen through the instruction of a good master, mountains to him are not mountains and waters are not waters; but after this when a person really attains to the abode of rest, mountains are once more mountains and waters are waters."

When we are children, if we are fortunate, we experience the unadulterated joy of contact with the world—digging for worms in the dirt, cupping our hands around tadpoles in shallow pools of water, chasing fireflies into a July night, blowing dandelion seeds to the wind. Mountains are mountains, rivers are rivers. The cutting and dicing, the naming and claiming, the classifying and dividing come later. That is when we see some larger purpose or use for the things of this world; we may even see that mountains and rivers are connected to all else, including ourselves. This is the stage when mountains are *not* mountains, rivers are not rivers. They are something else, serving some other end. We might even be tempted to think we know what they are truly good for. This is perhaps a necessary stage of growth but not a sufficient one. One cannot fully appreciate that which is only a subset of ego.

The profundity of the third stage, "the abode of rest," when mountains are mountains and rivers are rivers once again, is that it so resembles the first. But with a critical twist. Mother and son sit on the sand and watch the sun drag pomegranate juice below the horizon with similar awe. But one surrenders to mystery without choice; the other willingly gives herself to the mystery, releasing herself into the depths of what cannot be easily named.

The greatest pleasures in life are serendipitous moments, and serendipity, while often surprising, is not random—the more open one is, the more grateful, the more one stumbles on this good fortune. We are surrounded by *koans*. They don't come in words—the swish of the universe is felt in a place deeper than words can touch—and sometimes they slap us in the face.

There was once a disciple who traveled a great distance to ask his master about the truth of Buddhism. The response: "Why do you seek such a thing here? Why do you wander about, neglecting your own precious treasure at home?"

■ ■ ■

Clap! The head plunges into the water. A shiny skin, a vessel that water sculpted and filled, is held flopping by a strong lance-yellow beak. The bead of his eye is expressionless. The fish jerks again, then sags in an arc of silver, cradled by the heron's mouth, the river dripping from head and tail. The sky, full of the heron with the fish, is mirrored below the water's surface.

A Question of Monarchs

A line of people curves like the body of a giant caterpillar, twisting up the mountain into the endangered *oyamel* fir forest. Michoacán, Mexico—the cloud-misted overwintering site of the monarch butterfly. Near the middle of the caterpillar of people, I keep pace, thinking about Catholic *peregrinos* who journey on well-worn footpaths and roads, sometimes hundreds of miles, to reach a precious relic—a lock of hair from a holy person, an ankle bone from a saint, a mural of the Queen of Heaven, Guadalupe. Pilgrims with penitential motives as varied as the hearts that carry them bear sins of commission and omission, secrets kept close to the breast. They wander, as the word *peregrino* denotes, in search of healing, to destinations where the sacred is rumored to breach earthly realms.

The sacred breaches the earth at Michoacán, manifested in the bodies of millions of winged insects, each weighing about half a gram. Let those with eyes see. At Michoacán, you would be forgiven for thinking you'd stepped into a fairy tale—an alternate universe of ruddy orange-and-black grape clusters, dripping from conifer branches, enveloping whole trunks, monarchs piled upon monarchs piled upon monarchs. Pinch yourself to be certain of reality. The single monarch that lights up a midwestern garden in the heart of Chicago or a hedgerow in New England—multiply that by tens of millions of delicate flutterings.

But numbers alone don't account for complexity. Or mystery. Monarch migrations are sometimes described as a "biological phenomenon" to suggest the staggering power of their long-distance flights.[1] The high-altitude forests of Michoacán mark a single arc in a migratory cycle that spans four generations and covers thousands of air miles.

Between March and April, a collective quivering begins and a river of monarchs releases into the wind, flowing toward the interior of the United States. Some branch off, following the Gulf of Mexico to southern states, but the heart of the river pours toward the Midwest, where monarchs engage in quick romances during the warm summer months.

It is August as I write this, nearly time for the fall migration back to Mexico. The monarchs I see alighting upon stalks of goldenrod or flitting between asters' purple sunbursts are the great-grandchildren of those who poured from Mexico about six months ago. These intergenerational migrants are readying themselves for a two-thousand-mile return. Unlike their parents and grandparents, who lived fast and loved hard, these offspring survive up to eight months, with the majority of their energy funneled into completing the longest migration of any butterfly on the planet.

All this from an insect whose weight equals a dusting of sugar.

My first and only visit to Michoacán's Monarch Butterfly Biosphere Reserve was an optional field trip at an interdisciplinary humanities conference in nearby Morelia, Mexico. I didn't know much about monarchs at the time and could claim only the dimmest understanding of their migratory cycle. Recently, however, I learned about monarch life cycles and more, much more, at a symposium hosted at the Chicago Botanic Garden. I learned that the location of the monarchs' overwintering grounds weren't discovered by scientists until 1975. I learned about all things milkweed, the host plant on which eggs are laid and monarch caterpillars feast exclusively. I learned about habitat loss and concerns about the impacts of climate change on monarch survival. I learned about the uncanny intersection of corn and soy production in the United States and monarch breeding grounds. And I learned about the withering, widespread impacts of glyphosate, a Monsanto-manufactured herbicide introduced in 1996 and designed to kill all plant life that is not "Roundup Ready," Monsanto's name for its herbicide-resistant, genetically engineered crops. About three-quarters of all corn and nearly all soy production in the Midwest is now of the genetically engineered, glyphosate-tolerant variety. Milkweed, the sole food source for monarch larvae, is decidedly not herbicide-resistant corn or soy. By some estimates, milkweed habitat loss due to glyphosate application now exceeds one hundred million acres.

Together, these conditions add up to a troubling—nightmarish—scenario for monarchs. Their numbers always fluctuate, but for the past two decades, the downward trend has been unmistakable. A record low was reached when population estimates fell from one billion monarchs in the winter of 1996 and 1997 to thirty-three million monarchs in the winter of 2013 and 2014. At that point, the acreage covered by their colonies was 90 percent below the seventeen-year average.

The absolute extinction of monarchs is not in question. A western population calls California home; they reside year-round in Florida; they also travel to the Caribbean; and Australia, New Zealand, and Hawaii host introduced populations. But the beating heart of the migratory population that pumps through the heartland of North America is faltering. In this case, it is not the wholesale extinction of a single species that raises concern but an entire suite of ecological relationships whose pulse is flatlining.

At the symposium, I put the pieces together in my head about what the loss of a biological phenomenon might mean. I reached a simple conclusion: we don't have a monarch problem; we have an agricultural problem. We've sacrificed robust prairies and their plants to the gods of ethanol, hydrogenated oil, and high-fructose corn syrup.

A reasonable person might roll up her sleeves and say, "Got it. Looks like a campaign to eliminate glyphosate would help monarch populations rebound. We can do this. Same as getting lead out of gasoline, or banning the use of DDT for mosquito control. Sure, it took concerted efforts in public educational outreach, maybe a bestseller like *Silent Spring*, and facing down the money, power, and spin of giant industries, but persistence will win the day, truth will be spoken to power. Let's mobilize." So, I waited patiently for the Q&A and put a question to the experts about what kind of political pressure was needed to eliminate the use of glyphosate. The response was unsettling: "There are fifteen new 'stacked' GMO [genetically modified organism] plants awaiting EPA [Environmental Protection Agency] approval," the speaker replied, noting that corn and soy plants are becoming resistant to glyphosate treatments and agribusiness companies are lining up to fill the chemical gap. Lop the head from the hydra, and seven more appear.

By the end of the day, my own head was in my hands. The erasure of a biological phenomenon. A migration refined, over the course of thousands of years, by billions of tiny insects intent on completing their round in the cycle. Wiped clean.

Coal mines and canaries are notoriously poor companions. Monarchs, monocultures, and agricultural biocides are equally incompatible, and some persons have begun to designate monarchs as the canaries in the cornfield. Inasmuch as the decline of monarchs represents the general struggle of insect pollinators of all kinds—on top of legitimate concerns about the quality of agricultural products and to what nonfood ends those products are diverted—some persons are wary that monarch declines may portend "food-chain collapse."[2] A friend of mine who works on prairie conservation told me, "Their lack of food today is our lack of food tomorrow." Yet there's no sign of a cease-fire on the land.

At the conference, we were given the take-home message to plant more milkweed—then hope for the best.

After all the information we heard detailing the magnitude of the problems, the milkweed admonition set off alarm bells of futility in my head. Really? More milkweed? That's the advice on which Joe and Jill are supposed to hang their despair—planting a handful of milkweed plugs? This, after we were repeatedly told that one perfect storm of inclement weather in central Mexico—a cold rain followed by a quick freeze—could forever snip the cord of monarch migration?

I noticed a feeling rising from my belly up into my chest: grief. Many of us swim in a miasma of negative environmental news. Panicked headlines with exclamation points once moved newspapers; now they garner clicks and retweets. I have a hard time not taking the bait. *Shouldn't I be a well-informed person?* I think, as my index finger twitches above the touchpad. In a digital age, as never before, the world's problems are at our fingertips (or thumbs). I sometimes wonder if the human heart is built to sustain the tonnage of such knowledge.

Science may make us informed citizens; it will not help us deal with despair. So, how does someone grapple with the grim prospect of losing a migration of monarchs that may be ten thousand years old, that may reach back to the Pleistocene?

I'm not a psychologist. I'm not privy to the latest on productive grief processing. Nor am I an expert on human behavior. But I can point to the possible ways that people deal with structural problems too large for any individual:

Option A. The file 13 approach. *Throw that nasty problem in the trash bin of the brain and preserve a clean conscience. Delete. Delete. Delete.*

Option B. The "meh" approach. *Usually accompanied by a jaded shoulder shrug and outstretched palms, this shows awareness of a difficult situation . . . but, "meh," what can one person do about it?*

Option C. The ultimate-victor approach. *Also known as the "earth bats last" defense, this tactic appeals to geological timescales, noting correctly and perhaps with a twinkle of moralistic reproach that humans will eventually go the same way as the dinosaurs and receive their comeuppance . . . but it may be a long time coming.*

Option D. The pragmatic fatalist approach. *In the face of disaster, there is a saying (hadith) attributed to Muhammad that captures this well: "If the Day of Judgment erupts while you are planting a new tree, carry on and plant it." Keep calm and carry on, do the right thing, despite the circumstances.*

Option E. Weep. (It's okay to feel and feel deeply.) . . . *Then begin a journey toward healing.*

I've tried all of these options as coping strategies, consciously or unconsciously, at one time or another.

Awareness does not suffice; confession is not enough. Monarchs don't care if I'm sorry, if I feel they've been wronged, if I rage away at those who profit from producing poison. So the best coping strategy I know is to follow the example of monarchs and *peregrinos*. Begin the journey, keep the reminders in my heart, protect what is sacred, and maybe one further (and the largest) step: restore the sacred when and where I can.

One of the bits of swag I picked up at the conference was a single milkweed plant. It's now in my community garden plot.

I know it's not enough. But for me, the sprout serves as a living reminder. A gesture toward the sky. A holy relic in the ground.

The single milkweed recalls not my own pilgrimage to Michoacán but the now-perilous flight that monarchs must make. Their annual journey, thousands of years in the making, undertaken by billions of their ancestors, is a present uncertainty. Their landing pads, scattered and broken, tucked between biocidal farms and impermeable concrete, layer novel meanings on the term *refugia*. Their flight has become a flight from danger, a search for safe haven.

Many peoples from central Mexico have migrated to Chicago, seeking opportunity and safe haven. In these communities, monarchs are a potent symbol of the journey of generations. A mural by Hector Duarte graces a wall of the National Museum of Mexican Art in Pilsen, Chicago. The painting depicts a chain-link fence in the process of unraveling, silvered steel becoming black-and-orange wings,

lifting to the sky. A border loosening, a divide transforming into freedom. When I look at this image, monarchs become much more than a pretty thing gracing a garden. They become an awakening. The prospects of monarchs provide a barometer by which to measure our ability to welcome the itinerant who seeks refuge, the pilgrim looking for a place to alight and find food and warmth and life.[3]

Monarchs are speaking to us, if only by their dwindling numbers, about the health of our commonly shared habitat. Let us see if our respective journeys toward a life-giving future can be conjoined. The evidence will be found, in plain view, by the fluttering of delicate wings.

III

Conciliation

Coyote Creates New Paths

One day, Monarch was flitting between gardens when she spotted Coyote gnawing on the corner of a building. Curious, she flew closer to see what Coyote was up to this time.

"Coyote, why are you grinding your teeth against this thing? Do you think it is made of bone?" Monarch asked.

"It won't be much longer," Coyote bragged. "I'm almost done."

When she returned the next day, Monarch discovered Coyote in the same place, still chewing. "Coyote, you must stop. Can't you see this is not working? Your teeth are getting dull. You have to go around this thing. You cannot go through," said Monarch.

Exhausted, Coyote stopped chewing and sat on his haunches. "You may be right," he conceded, realizing he had made no progress at all.

The two rested together for some time, looking up at the sky, saying little. Then, Coyote's ears pricked up. He ran to the river with Monarch close behind. When he reached the bank, he began scooping mud up in his paws, and throwing it wildly about in all directions. Monarch dodged this way and that, utterly baffled. Coyote worked hard, into the night and beyond, and the longer he worked, the crazier his throws became, until finally he threw so hard he came apart completely, his tail going one direction, his arm another, his leg flying off to the south, another to the north, and his head landing with a thump where he once stood.

A horrified Monarch flew to the other animals with the sad news. When the animal-people who attended Coyote's council in the woods found out what happened to Coyote, they came from all directions to see. Gathering in a circle around what remained of Coyote, they mourned their friend's condition.

"He did try, in his own way," said Mole, whiskers vibrating.

"Yes," agreed Heron, slowly shaking his yellow beak, "he was foolish and brave."

Then, one by one, the animals brought all of Coyote's parts and placed them in the center of the circle. They stared, not knowing what to do except to be troubled. Fox, a quick thinker who knows a bit of magic, strode up behind the mournful animals. When he peeked between Heron's legs and saw Coyote, scattered about like a jigsaw puzzle, he felt pity for his cousin.

"Why do you stand around, looking so glum?" spoke Fox to the others. "This is a small matter." And, so saying, he whispered some words and rubbed ash on Coyote's bones. When Fox had circled the bones five times, they began to quiver against the ground, sliding closer, then locking together, until Coyote was all in one piece again.

Coyote stood up and blinked, surprised to see the other animals gathered around him. His vision was still clouded. Squinting, he looked down at his body.

"What happened to my fur?" he demanded, for his fur was stained black in various places because of the ash Fox used, especially the tip of his tail.

"Don't worry about your fur," said Heron. "You should be saying thank you. Fox put you back together."

"And all of us helped," squeaked Mole.

"Did I do it?" Coyote asked.

"What do you mean, 'Did I do it?'" Heron scolded. "Break yourself to pieces because of your wild ideas? Yes, I just told you, and we put you back together. How could you do that yourself?"

"No, no, no," Coyote scoffed. "Did I build the paths between the buildings? Did the mud work its power?"

And, sure enough, looking around, the animals saw that Coyote had done something great indeed, and they were no longer angry with him. This is how new paths came to the city.

Shagbark Thoughts

I've lately been visiting a neighborhood pocket park in the early morning, book in hand. An underused triangle of grass, the park contains a bench tucked haphazardly into the far corner like a city planner's afterthought. I cozy into the bench on cold days, nestled in the lingering warmth of a few layers of clothing and a down jacket. Until the first cup of coffee enters my bloodstream, the cold and lack of caffeine conspire to slow my thoughts, my reading pace consistent with the sap flowing through the limbs of a nearby shagbark hickory. Ideas percolate, bubbling to the surface.

I think on these small patches wedged into neighborhoods. The park's trees are decades older, and sometimes hundreds of years older, than the unfamiliar newcomers in surrounding yards. Witnesses to so much change: watching the homes go up, the prairie become lawn, the footpaths become sidewalks, the compacted dirt become streets resisting their restless roots. Still, they watch, memory keepers, more valuable now because of their thinned numbers. In summer, the squirrels rejoice in their seed, practicing acrobatics above. I rejoice in their shade, focusing my attention below. In winter, stripped to their bones, they rest from their labors, waiting for the earth to tilt back toward the sun. Dormant, they still exude presence, still hold the space, still keep memory alive.

Just a little triangle of breath between concrete. But a place where children build snowmen and arrange sticks into imaginary camp-fires, listen to birdsong and cradle pill bugs in the cups of their hands. A place to reflect on this landscape's history, a place to expand that history into the future.

I look upward at the lengthy telltale strips peeling away from the trunk that give the shagbark its name. Lean down, pick up a seed husk. Turn the smooth, ovoid wood in my palm. Run my finger along the curve of the pit. I am thinking of the gift of small things—patches of grass between buildings, gardens in vacant lots, even balconies that open up to the sky. I am thinking about the importance, the ne-cessity, of these small encounters for a reimagined future, of tying the memories of the past to a fuller present. I am thinking of a little African American girl and her fondness for pigeons.

Sherry Williams is now a woman on a mission, a person with an interest in recovering African American history and genealogy in Chicago. She is also an amateur birder. She can trace this interest to a humble balcony from her childhood and to the pigeons she be-friended there.

Same as a lot of Chicago residents, Sherry lived in a walk-up apart-ment without a yard in the front or back. She didn't have any pets, so her connection to nature came through birds that were present year-round: pigeons. When her mom later moved the family to a house, she became especially concerned over her daughter's pigeon-feeding compulsion. She issued stern warnings to Sherry that squirrels, at-tracted by the food bits, were bound to wreak havoc in the attic. Sherry remained undaunted; she chuckles softly when she recalls for me how she kept "sneaking" food to the birds despite her mom's protests.

Pigeons are the opposite of a conservation concern, innumerable in cities throughout the world. Andreas, a German friend of mine, tells me that disparaging pigeons transcends international boundaries. In Germany, they are known as *Luftratten*. Air rats. Common, annoying, ordinary, ignored—pigeons may be the bird most taken for granted in urban areas. Despite their numbers, ironically, pigeons are almost invisible.

Not for Sherry. Pigeons brought her peace, she says. When I ask her why she persisted in her bird-feeding rebellion, she responds: "My entire life it's just something I enjoyed. Of course, it does bring peace, [and] it's such a joy to hear their calls. That of course over the screeching that you would hear when you live on a major street." Small places, small encounters. We start where we are, with what is available to us. An iridescent bird who has learned to seek bread from human hands is not a bad place to begin.

Which reminds me of the wise words of writer and lepidopterist Robert Michael Pyle. There's a moving chapter entitled "The Extinction of Experience" in Pyle's memoir of childhood exploration, *The Thunder Tree*. There he notes that when people think of extinction, they often think of rare species—Javan rhinos and Bengal tigers. Animals that are large or furry, or a combination of both, draw the lion's share of our attention. As important as big, charismatic species are, Pyle sets his sights on something closer to home. He expresses concern over a potentially more devastating loss, "the extinction of experience." This type of extinction is more subtle, occurring at the scale of the neighborhood, and therefore less appreciated and harder to detect. Rarely is ink spent on headlines about this type of extinction. One season a copper-colored butterfly flutters by, the next year it is gone, cool concrete substituted for purple coneflower.

These small disappearances could be characterized as lesser losses. After all, in Chicago there are the forest preserves, many miles of lakeshore, innumerable municipal parks. Even if we get to those places only once in a while, the Discovery Channel or Animal Planet or Nat Geo Wild allows the exotic to burst into our living rooms. For Pyle, this way of thinking reflects the slow erosion of intimacy with so-called throwaway landscapes.

People who are crazy about nature almost always have a story about a throwaway landscape—a place that others have difficulty seeing as valuable, but that for them defined freedom and solace during their formative years. "Ignominious, degraded, forgotten places that we have discarded, which serve nonetheless as habitats for a broad array of adaptable plants and animals" are the secret gardens of imagination for children who later become adults that care about the natural world. Muddy stream embankments, overgrown housing lots, abandoned fields rutted with dirt-bike tracks—these would never be mistaken for the "jewels" of any national park or wilderness area. According to Pyle, however, in some ways they serve more formative roles. "Along with the nature centers, parks, and preserves," he writes, "we would do well to maintain a modicum of open space with no rule but common courtesy, no sign besides animal tracks. For these purposes, nothing serves better than the hand-me-down habitats that lie somewhere between formal protection and development."[1]

Hand-me-down habitats, throwaway landscapes, and second-hand lands do not merely spark a sense of adventure. They provide us close-to-home access to more-than-human worlds, a way to learn who we are in relation to many other species with whom we share a common patch of earth. They unlock portals of imagination, worm tunnels that blast us into new galaxies. They gift us with familiar common ground for developing caring relationships.

It's not that exotic animals in foreign lands aren't worthy of concern. In the grand scheme of planetary complexity, they matter deeply. But contact, *real* contact, leads to relationship. Borrowing a page from Leopold, who wrote, "We only grieve for what we know," Pyle lays his point out plainly: "People who care conserve; people who don't know don't care. What is the extinction of the condor to a child who has never known a wren?"[2]

Or even less exotic than a wren—in Sherry's case, the humble pigeon.

A red-tailed hawk throws a shadow on a solitary cardinal and a congregation of pigeons, chickadees, and monk parakeets. Nobody seems bothered. Underneath the hawk's watchful gaze, I step into an unlikely bird oasis near 111th Street and Cottage Grove Avenue in South Chicago.

I'm certain this avian gathering did not constitute a part of George Pullman's plans. George Pullman, train entrepreneur and capitalist utopian, moved to Chicago in 1863. He had an ambitious vision: create a self-contained manufacturing city. In four years (1880–1884), four thousand acres of marsh and prairie sprouted a complete town with its own market, sanitation, entertainments, and, of course, workforce. It was even voted "the world's most perfect town" by one international body in 1896.

I walk through the streets and between filigreed buildings, and my imagination easily drifts backward in time. Nostalgia clings to the bricks. The local McDonald's displays vintage historical photos. The buildings show their age and their tarnished dreams, frayed but beautiful. They seem to be waiting.

The now-derelict Pullman factory is where Sherry, the pigeon lover, had a vision.

Sherry became familiar with local birds on her balcony, but Chicago as a whole serves as a kind of front porch: a major migratory flyway for birds, who use the area as a layover stop on their seasonal journeys. And as Sherry is quick to point out, Chicago has also been a prominent place for the migration of people. Moving, settling, seeking welcoming habitats and opportunities, connecting distant places of concern—humans and birds move across landscapes in the hopes of safe harbor.

From the late nineteenth century until the mid-twentieth century, thousands of African American people were drawn to Chicago by employment opportunities, hoping to escape racial barriers in the South. This Great Migration figured prominently in Pullman, where the first all–African American union (the Brotherhood of Sleeping Car Porters) was established.

These two streams of interest merged for Sherry behind the defunct Pullman factory clock tower. As she thought about human and bird migration, a connection began to crystallize. Another pigeon, this one a more ethereal dove, provided the inspiration; as Sherry explains, "The Holy Spirit has a way of inclining ideas to you."

For Sherry, constructing a bird oasis became a way to better understand and welcome avian migrants that use the area, as well as highlight the migration stories of people in the community. With the help of the Illinois Audubon Society, young people who needed service hours to fulfill Chicago Public School requirements, and curious neighbors, Sherry began hosting guided tours about once a month. The transformation of the site included repurposing a mowed path that was once used for security checks of the Pullman property into a footpath for exploration; taking out invasive shrubs and replacing them with native plants and trees attractive to birds; adding feeders, baths, and birdhouses; and creating a central picnic area.

Though all ages of people go on the walkabouts, Sherry is particularly concerned with young people. Several times she mentioned gun violence as an ongoing problem that the community faces. As an increasing number of birds have discovered the site (a recent winter count resulted in sixty-three species), "slowly but surely young people have become attached to the project." Students who came initially to accrue community service hours now learn about African American migration parallel to the migration of birds in Chicago. They've made this the core of what they discuss, Sherry says. She sees these experiences and conversations as a way for them to "become that next generation of stewards." During months with favorable weather, signage on the walking path displays narratives of human immigration history as well as information about bird migration, building on the conversations Sherry has fostered around the oasis.

In the midst of so much movement, the oasis is aptly named. A place to seek refreshment. A place to stop and reflect. A place to think about people and birds—and how our stories may bear affinities. For Sherry, a connection to the common pigeon unfolded into a larger vision. Out of disrepair, new habitat has begun chirping. A larger conversation between people and place has begun.

■ ■ ■

I'm not a great gambler. Whatever attracts lady luck, I don't seem to possess it. I joined the Cub Scouts in grade school, and the memory of a monthly gathering of various Cub Scout packs still feels fresh. I recall only one thing about those monthly meetings: the cake raffle. For fifty cents a ticket, you could buy a chance to win a cake. My mom always spared me one dollar. Two tickets. Never mind that some kids clutched a string of tickets reaching from fist to foot. All I needed was for *my* number to be called.

On a typical night, ten to fifteen cakes populated the table. As numbers were announced and kids rose from their chairs to select the most delicious, icing-laden sugary delicacies on the table, I hoped until the final number's hollow echo against the walls of the rented church auditorium that my ticket was a winner. I would have settled for the last lonely, dry, poppy-seed-encrusted Bundt cake. But never in my four years as a Scout did I cash in a ticket.

I presumed I was cursed. So it was no small relief when, in adulthood, I finally won a lottery. A garden lottery, for a plot in one of our city's community gardens. True, it wasn't a cake, and I actually had to pay for the privilege of renting the plot, but it brought consolation to the child in a Scout uniform who once pined for triple layers of chocolate.

I'd had a backyard garden long before, but now that we lived in a city apartment, I missed tending young seedlings, digging up a few scrawny vegetables and scrappy herbs for the table, and, mainly, the pretense to be out under a warm sun with soil-stained hands. When I won the lottery, big ideas were set in motion. I thought about space optimization, accounting for shade and sunlight. I contemplated accessible pathways to plants with a carefully hand-drawn map. I participated in a local seed swap, ordered exotic-sounding heirlooms from catalogs, and put a solid set of blisters on my fingers while turning the soil with a borrowed hoe and shovel.

By midsummer, though, it was clear I wasn't going to reap anything close to the bounty I imagined. Only a smattering of plants hung on. Then I started seeing the rabbits. Soon after, the hoofprints of deer. My raspberries went to the birds; my tomatoes to the chipmunks. When a plant appeared to be near fruiting, I began a race against the clock, and the other neighborhood animals were more frequent garden visitors than I. Finally I began to call my garden what it was: a wildlife donation.

I learned to come to terms with this reality. In fact, I began to think about how I could be a more proactive donor.

Over the course of the next few years, I lessened the square footage for vegetables and increased the number of native plants that served as hosts for butterflies and offered nectar to bees. I managed to pull quite a few carrots and pluck the occasional pepper for my own family, but I also began to think of the garden's broader bounty, a dot of habitat within the larger green matrix of the city. I noticed my view of lawns and vacant lots and road medians changing too. Wherever I turned my gaze I saw life-giving potential.

I'm not the only one. Robert Nevel acquired this vision long before I did and has been putting it to use on larger scales than a single plot in a community garden. Nevel is the president of KAM Isaiah Israel (KAMII), the oldest Jewish congregation in the Midwest. Trained as an architect, he one day looked at the grounds surrounding KAMII's historical central-dome synagogue on Chicago's South Side, and, where others saw manicured grass, if they noticed the lawn at all, he saw opportunity.

It's typical, Nevel tells me, for congregants to view a house of worship as a mass in a sea of grass, treating the surrounding lawn as leftover space. For most of the synagogue's history, the grounds were indeed an afterthought, some *thing* to walk through on one's way to what was truly important and unbound by earthly soil.

As a congregation in the Reform Jewish movement, KAMII holds social justice as a core part of its identity, so Nevel believed there was an open opportunity to connect the dots between social and environmental justice, with the lawn as the canvas. He knew, however, that tearing up the lawn was going to be a tough sell. Noting the devotion of Americans to their lawns, he wryly remarked, "In its own odd and ironic way, lawns have become a sacred space." The solution would have to include easing his fellow congregants into an alternative perception.

He proposed designs for a hexagonal garden in the front lawn of the synagogue, in the shape of a six-pointed Star of David. Each of the star's points would grow food; the negative space would remain lawn. The proposal was approved. Work began in 2009. The Star of David remains, now anchoring a much larger transformation: gardens surrounding the synagogue have doubled every year, and the former lawn on the 2,500-square-foot property is hard to find between the tomatoes.

Further connections, both religious and secular, came quickly, and the evidence radiates into the larger community. Some of the most remarkable stories involve the way in which the gardens mediate interfaith relationships and understanding. Nevel recalls a moment when he watched Muslim children attentively listening to an eighty-year-old member of the congregation read a book about growing carrots. "Food and care for the earth is common to all of us," he explains.

In many ways, gardens are the epitome of domestication. But in a city, where concrete and manicured patches of turfgrass rule, a garden is a step toward rewilding, a welcome call to the agency of other beings, from microscopic soil organisms to four-winged pollinators. The gardens at KAMII serve a very practical purpose—feeding people in the community. They also represent lifelines that reach still further, woven together by the aerial surveys of goldfinches, foraging bumblebees, soil organisms, and a hundred different gleaming beetles. The value of such vital places—one might even say their sacredness—is that they add a measure of wildness to the city's fabric. The garden is not solely an invitation to other creatures; it is an invitation to remember how the human creature is connected to soil, water, air, and nonhuman neighbors who share our urban spaces.

"I think that anybody who works in this program is changed when they work in it," Nevel reflects. "They see their responsibilities to each other and the land differently." Another way of putting this: much more than the lawn of a synagogue has been transformed. People go from living *on* a landscape to living *within* one, as participants in the greater well-being of their community. Nevel summarizes this for me:

"They see it under their fingernails. They see it on their knees. They feel it in their back. They can see, and feel, and taste the difference that they're making."

When I visited KAMII, I explored both the buildings and the gardens that surround it. It occurred to me that while a building may unite people in common purpose and belief, the grounds were a celebration of the world to which we, all of us, are connected. The big congregation of life. Perhaps gardens like those at this synagogue reveal how the wild and sacred can be cultivated. Perhaps the two words, wild and sacred, are closer in meaning than most of us imagine. Perhaps they both can be seen under one's fingernails.

· · ·

Balconies, backyards, gardens. Thinking about how rewildings of the landscape can be scaled up and out, I went to visit another garden—a very large urban garden—known as the Perry Avenue Community Farm.

On what happened to be his sixty-sixth birthday, I met with Orrin Williams (no relation to Sherry), the founder of the Center for Urban Transformation, a co-leader of the Perry Avenue Community Farm, and a person interested in a revolution on Chicago's South Side. I first became aware of Orrin while attending a conference at Chicago State University and was immediately impressed by his casual candor. His salt-and-pepper dreadlocks hint of the radical, the lines in the corners of his eyes speak to an earned perspective. I discovered that we shared an interest in urban agriculture, particularly the way it provides a medium for social and ecological exchange, so I sought him out to hear more.

This led me to the Perry Avenue Community Farm, a two-acre chunk of rectangular land situated on a former school parking lot in Englewood, Chicago. When the neighborhood of Englewood makes the news, it's typically not a good thing. The area is one known for its gang violence, still reeling from the institutional racism that stalled business prospects, gutted its local economy, and prompted white flight to the suburbs.

Orrin has witnessed it all. He came of age when a black middle class in Chicago was more robust and fondly recalls how, as a child, he could buy live chickens a couple of blocks away from his home. There was an A&P supermarket. There were local grocery stores. At Sixty-Third and Halsted were theaters, banks, and the second-largest economy in the state, behind only downtown Chicago. To get what your family needed, Orrin says you didn't need to leave Englewood: "Your mom would hand you a list and off you'd go." But he also experienced the manufacturing decline in Chicago in the mid-1950s and watched segregation draw the boundary lines around his neighborhood's prospects, including the construction of an expressway that bypassed Englewood and became a catalyst for suburbanization. He tells me there were no vacant lots in his neighborhood growing up. Now, there are no houses standing on his childhood block. He's interested in changing that.

After pulling up to the site and having a brief look around, I follow Orrin inside a once-vacant home that has been refurbished into a beautiful community center. A stunning mural with a collage of colorful visages of African Americans stretches across the entire north side of the two-story building. Inside, an aquaponic system hums in the background; photographs of people holding radishes, watermelons, and chard line the walls; a dry-erase board, with various brainstorming notes scrawled across it, such as "[SOLUTION-]/ARIES" and "Consumer Re-connect," dominates one corner. Orrin explains that this is the "think-do house," a place of community reflection, strategy, and action.

I had come to talk to Orrin about how food can connect us to place and to one another, how it can create a sense of rootedness and home. But our conversation soon veers into many interrelated topics. Two mind-expanding events in Orrin's life fed directly into the creation of Perry Avenue Community Farm. The first happened during the Vietnam War, when he was stationed in Thailand. It was there, half a world away from Chicago, that he saw alternatives for food. As he puts it, his little urbanized, American brain exploded "from the first moment I stepped off the plane. It smelled different, it looked different, the people looked different. You didn't have a car. You didn't have a furnace. You didn't have air-conditioning." Thailand was awash in green vegetation, rice paddies, farms, and villages with ponds full of fish. Mangoes were as common as acorns in Chicago parks. Orrin's consciousness around food expanded as he connected to a new pace and rhythm of life, sometimes plucking breakfast straight from the branches of backyard trees. A number of years later, in a second revelatory event, he became aware of his own family's ancestral connection to farming as well as a broader movement to secure and reclaim black farmland. He recalls visiting Pembroke, Illinois, sixty-five miles south of Chicago, where a community of black farmers persists to this day. His eyes opened wide when he first saw this community, stunned as he was by seeing black people farming, and his mind filled with possibilities.

Although food provides an intergenerational pathway for people to reconnect to place, Orrin notes that local provisioning is merely one part of re-creating twenty-first-century cities. Communities also must reckon with the need for better-quality housing and offer viable local commercial and entrepreneurial opportunities. Orrin tells me it is critical to put all the pieces together—the social, the economic, the ecological—integrated within a dynamic community. This includes reviving traditional arts and music. As he puts it: "There's no word for art in a lot of societies. It's lifeways. It's all part of an aesthetic. Schools don't want to do it. We want to do it right here." This vision for lifeways that nurture community could be especially important for neighborhoods such as Englewood, where unemployment rates hover around 50 percent. "What the communities are notorious for is really just a small part of what goes on," Orrin says, referring to the periodic and all-too-common shootings on the South Side of Chicago, but "if you don't have life for people to do, then other things come to them."

When I ask Orrin if healing in the community applies to non-human Chicago residents, he points toward the farm and chuckles, "We try to grow enough to make sure they have some too." He then becomes philosophical: "The planet is not here just for humans, right? It's here for all kinds of sentient beings. And we have to be respectful to that. I think their *being* is important as any human *being*, if you will. So we gotta try to do things in ways to heal the planet and allow them to thrive as well. I look at communities as ecosystems; we are part of the earth ecosystem, we are part of the cosmos." From the planetary to the microscopic, as Orrin puts it, "I think people are connected to place in ways that we don't even recognize. We're connected to microorganisms, I mean they're all over us. The health of your gut determines how healthy you are. The whole thing is it's all one thing. It's all about oneness. That's all there is to it. Now whether we connect to that, recognize that, whether we're conscious of it, I think that's what makes the difference."

Orrin leans back in his chair, pulls out a smartphone, and exclaims, "We get dazzled by all this bullshit." He lays the phone down and puts an index finger to the side of his head. "The revolution happens here," he says as he taps his temple.

Orrin's right, of course. Ideas—especially the collective power of shared ideas—give us an architecture for the future. This is "think-do" work: the revolution happens in the mind *and* in the places we live. "People say, 'I think I'm going to leave Chicago,' but then they say, 'Where am I going to go?' The problems are everywhere." As an example, he asks me, "Why are we importing California's water? I don't care if you call it lettuce—it's water. They don't have any water, but they are exporting water to Chicago. . . . What's the interconnection between those communities and this community? What's the interconnection between our city and other places on the planet? Whether it's food, whether it's art, whether it's healthy children, I don't see the disconnection between any of that."

Local food is a place to begin, a strand to pull on that connects us to others on the landscape, human and nonhuman. Orrin sees an opportunity in the midst of the vacant land for various uses: "Some of these spaces, even in places like Englewood—I don't want to see farms and I don't want to see houses on them. Can they become bird sanctuaries or habitat for some of these creatures? The block I grew up on, I was over there visiting my buddy, and he said, 'See all that stuff over there.' And there was a bunch of trees and grasses, and he said, 'Man, at night, all kind of shit comes out of the trees,' like raccoons and opossums. Maybe we need to create those islands and let 'em be. Maybe we don't want to rebuild every neighborhood. Maybe we want to let it go wild, if you will."

As Orrin speaks, I reflect on the disasters that have struck major cities recently. The economic collapse of Detroit, the devastation Katrina wrought in New Orleans, and Chicago's patchwork of affluence and poverty. Cities are cauldrons, shifting brews of destruction and opportunity of our own making. They are also places of experimentation. "As we approach the point that 50 percent of people live in cities, I feel a transition," Orrin reflects, "I don't think we know but I think there's some wonderful possibilities. The planet is going to be here. We're not destroying the earth; we're destroying the possibility of being here in this form. Until the asteroid hits, we need to fix what we do here. The experiment has begun. We're part of that emerging understanding of what we have to do to be here."

An experiment has indeed begun. People like Orrin are rethinking the city from an ecological point of view. They've lived through transitions, watched urban neighborhoods vacated and reborn. Speaking with Orrin, I can't help but get the sense that another change is coming, a think-do effort to reweave our cities into something better for humans and nonhumans alike along a wild continuum.

. . .

Vital places draw our attention to the life that pulses through our everyday worlds. In urban areas, these places help us rethink what our responsibilities to nature might be by reframing our ideas about *where* nature is. A balcony, a backyard, a garden, a city, a bioregion. Awareness doesn't always follow in that order, radiating outward like the concentric circles of a pebble splash, but an initial encounter with other species can grow into a wider effort to connect the dots. It certainly has for Sherry, Robert, and Orrin.

I'm under the shagbark tree again, cozied into the corner bench. A slight breeze pinches my cheeks. I put my book down and close my eyes, thinking about the strange and wonderful beings that share this little nook with me, thinking about the small and vital gifts of green space, known or forgotten, scattered about the city. In *The Thunder Tree*, Pyle reminds us that the way to caring about other species begins on the footpaths close to home. "What, to a curious kid, is less vacant than a vacant lot? Less wasted than waste ground?" he asks. These places cradle small discoveries, nurturing moments of delight, and "somewhere beyond delight lies enlightenment."[3]

I stare through the branches of the shagbark, delighted by what this tiny triangle of green park space brings to both the shagbark and me.

Vole-a-Thon

Reintroduction biologist Allison Sacerdote-Velat cinches a white mesh bag around the dark rectangular opening of an eight-inch-long box. Inside, a faint scritch-scratching can be heard of tiny toes against metal. She tilts the box downward. Nothing happens. She increases the angle. Still nothing. She gives the box a gentle shake—an agitated *squeeek*, akin to a bird chirp, echoes outward, and a small lump falls headlong into the waiting softness of the mesh. Allison puts the box on the ground, closes the bag, and hooks a scale on the top. Lifting it to eye level, she peers into the mesh. "Hi, little guy."

A few quick scribbles on a pad of paper, and then Allison reaches into the mesh and withdraws a ball of fur, pinched between her thumb and two fingers. Dangling in the air, paws extended, pink nose twitching in front of a pair of glassy black eyes: another vole.

I am tagging along with Allison, a scientist with the Urban Wildlife Institute, to get a better understanding of the labor that goes into reintroduction biology.[1] She along with others—today, intern Rachael Pahl, whom we long ago lost in the tall prairie grasses—monitor several "species of concern" in the Chicagoland region, including wood frogs, spotted salamanders, smooth green snakes, meadow jumping mice, and the least weasel. Sometimes the work involves releasing captive-bred species into the wild, sometimes translocating individuals to boost an isolated population's numbers or to ensure they have more viable habitat—always, at least when the funding is available, reintroduction biology involves monitoring. This research requires a long view, asking how different management techniques and habitat restoration practices have an impact on species' populations over time.

Voles, such as the one Allison holds in her hand, are not a species whose populations are in any danger of extinction. But knowing what they are up to and keeping track of their numbers is one puzzle piece in the prairie jigsaw.

I know a prairie steward who refers to voles as "the hamburger of the prairie." Almost every carnivore—raptors, snakes, coyotes, foxes, to name a few—eats them. They are year-round selections in the prairie buffet. Voles don't hibernate during winter, and as Allison notes, "They never stop eating; they never stop breeding." Which can be a problem if you're trying to restore a prairie. The collective impact of their voracious appetites on plant roots can make life difficult for rare and threatened flora.

After examining the vole for evidence of a botfly infection—a fairly heinous parasitic ailment to which rodents are susceptible—Allison picks up an instrument that looks like an oversized pair of tweezers and clips a small metal tag to the vole's ear. The vole emits a tiny squeak of protest, and she sets the four-inch creature in a patch of dry grass. He pauses a moment, back on solid ground, then darts away, a blur of brown fur.

We are traversing a parcel of land called Rollins Savanna, a fairly high-quality mix of tallgrass prairie and open woodland. It's the middle of September, but most of the waist- to shoulder-high vegetation hangs on to various shades of green, punctuated by the lemon-yellow blooms of feathery goldenrod and black-eyed Susans. We spend most of the day out in the open, occasionally venturing into the dappled shade of bur oak trees. The air is muggy, and I'm sweating by midmorning. Despite the heat, I opt to roll down the sleeves of my shirt in order to keep the mosquitos at bay.

While we walk, I frequently run the palm of my hand over the tall grasses, letting the seed heads tickle my flesh, breathing deeply of the late summer sunshine. This site is among Allison's favorites, in part because it has transformed so much in a decade. It was once just a thicket of shrubs, she tells me, difficult to pick one's way through. Buckthorn and honeysuckle, invasive species that aggressively shoulder out local plant diversity, pose a persistent problem on many remnant prairies in the area. Rollins Savanna is no exception. Volunteer restoration groups and county forest preserve employees have done—and continue to do—a tremendous amount of restoration work in these areas, making them more viable for the kinds of rare and threatened species that Allison monitors.

I am accompanying Allison on this site visit because I want an expanded sense of the challenges faced by urban wildlife. Not all, not even most, animals are as adaptable as coyotes—who are grouped into a category of creatures called synanthropes, meaning they are not terribly bothered by human activity or may even be drawn to places of human habitation. The animals Allison studies are those with more specialized needs.

Take the eastern massasauga rattlesnake as an example, a locally extirpated species who not only requires sizable connected habitats but also overwinters in crayfish burrows and so depends on particular hydrological features. Their numbers have been so reduced that, for the moment, they exist only in captivity. Another example is the least weasel, the smallest mammalian carnivore in the world. Less than a foot long and similar in shape to a submarine sandwich—if that sandwich had a handsome set of whiskers—the least weasel has reddish-brown fur, four stumpy legs, and a cream-colored underbelly. Least weasels don't do well with concrete, and their dwindling numbers are a concern for scientists like Allison. The only captures for monitoring purposes so far have occurred in the suburbs near more rural lands.

As I look around me at the positively bucolic landscape, it's easy to think *country* not city, but Allison reminds me that the third largest mall in Illinois, Gurnee Mills, is just down the road. There are also seven hundred thousand people who live in the modestly sized Lake County, where Rollins Savanna is located. The commute to Chicago is a smidge over an hour, depending on traffic. "These are islands of habitat in a sea of development," Allison tells me. Rollins Savanna provides a good example of the impacts of the urban gradient, the way the city reaches into more rural areas via roads and infrastructure. The presence of the city manifests here in the form of habitat fragmentation, as chunks of land are divided from one another by cement and asphalt, stranding creatures on pavement-locked islands. The bigger critters aren't the only ones who struggle.

"Hi, buddy," Allison says into another box trap. When she's done with her protocol, she places the vole on the ground along with the seeds that lured him into the box. Unperturbed, the vole continues to nibble the victuals while she makes further notes. "I've always been drawn to the underdog species," she says. "I love vermin. Snakes, weasels, species that people have worked really hard to eradicate." Her career has taken her all over the country—desert tortoises in Utah, sharp-tailed sparrows in Maine, mountain lions in Idaho, silky-flycatchers in Las Vegas. I close my eyes and try to imagine what a silky-flycatcher looks like.

Roads present the greatest problem for the prairie-adapted species she studies in this area. Crossing structures for wildlife, which are becoming more prevalent in some parts of the country, are still few and far between. Even if they were built here, they might not be much help to snakes and smaller mammals, who tend to cross a road wherever they happen to be, with or without the benefit of a crossing structure to lessen the peril. Allison gestures toward Route 83, which marks the southwestern border of Rollins Savanna. "I wouldn't want to cross that road as a small mammal. I don't even like driving on it because people drive so fast." Without a massive restructuring of roadways and, more realistically, greater forethought about how and where to build them, the species Allison monitors face persistent challenges. "The more bound species are, the more we need to understand how we can keep them going, or *if* we can keep them going, or if we *should* keep them going."

At the next box trap, she gingerly handles a vole with a botfly, one of many we will find today. It's heartening to meet scientists such as Allison, the kind of person who stopped our car as we drove between sites to move a leopard frog off the road and away from potential harm. She talks to the voles while she gives them their ear tags— from friendly greetings ("Hey there, little guy.") to sympathetic cooings if she finds evidence of botfly infection ("Ahh, poor little thing. That looks so painful.").

Much of the day is filled with the sound of pant legs swishing between prairie grasses, watching butterflies move with the wind over the spiky tops of rattlesnake master seed heads, and discussing the behaviors and distinctive qualities of other animals. I learn that masked shrews smell like old socks; voles like a musty attic; weasels like burnt mustard mixed with sulfur; and jumping mice, who spend a lot of time grooming, have a clean—or clean*er*—scent than most.

I keep hoping we will find a jumping mouse in one of the box traps, but our day turns out to be a veritable vole-a-thon, revealing that the hamburger of the prairie is thriving. That's good for some species, not as good for others. Good for least weasels, a species considered "of greatest conservation need," for whom voles are a primary source of food. Allison tells me that sometimes when weasels find a vole burrow, they eat the vole and then line the burrow with their fur to comfy up their new domiciles. "Using the whole buffalo," I remark. But all the voles we find in the traps may not be so good for jumping mice, whom voles tend to outcompete, in part because jumping mice aren't nearly the reproductive titans that voles are.

Just because we don't find any jumping mice on this particular survey doesn't mean they aren't here, Allison assures me. It's a complicated world on the prairie, and the mice might be at a point in their life cycles not conducive to live trapping. No need to raise the alarm.

Monitoring requires patience. If Allison and her team consistently find a lack of jumping mice at this site, they might recommend a supplemental reintroduction. I ask if she sometimes feels as though she and her colleagues might be putting off the inevitable with these kinds of supplementations. That's what the monitoring and population models are for, she responds.

I ask Allison to speculate a little. How would she design this landscape if she could call the shots? More wildlife crossing structures would help, she replies, but because each species has such different needs, it's hard for her to pick one thing. After some thought, she seems to settle on a recommendation: "I suppose if you're going to build a road, think about the path of least disturbance." The species she studies simply need enough connected habitat to thrive. Natural and human-cut corridors—rivers, land beneath power lines, and especially railroad right of ways—serve as major migration paths between otherwise isolated animal populations. Smooth green snakes, for example, are consistently found near railroad tracks, because they depend on these links between scattered islands of prairie. Probably the best thing for all these prairie-dependent creatures, she adds, occurs when public conservation agencies acquire these lands before roads are built.

Allison has been a field biologist long enough not to be starry-eyed. There's a reason the species she studies are underdogs. Most people continue to believe that open land is wasted land, a blank spot of weedy nothingness, hosting inconsequential creatures. I've met people in the Chicago region who are simply in love with prairie landscapes, but I also know others who struggle to perceive a landscape's beauty or value if it doesn't include a forest or a mountain. And if a natural area is close to the city, many people may not think of it as a landscape at all.

I wonder if voles could be prairie ambassadors, helping city folk see the urban gradient with different eyes, as a latticework of prairie grasses, wetlands, riverways, and forests that can be rewoven for the benefit of us all. A lot can be learned from the hamburger of the prairie, a creature whose life is connected to so many others.

"There you go, buddy," Allison says to the final vole we encounter on our transect. She kneels to tip the mesh toward the ground. With a skittering of legs the vole scampers away, a mobile source of information about the prairie landscape on four tiny paws. *Good luck*, I say, and Allison smiles as she scribbles a last note for the day.

Desire Lines

In this Route we see only vast Meadows, with little clusters of Trees here and there, which seem to have been planted by the Hand; the Grass grows so high in them, that one might lose one's self amongst it; but everywhere we meet with Paths that are as beaten as they can be in the most populous Countries; yet nothing passes through them but Buffaloes.

—PIERRE FRANÇOIS XAVIER DE CHARLEVOIX, French Jesuit explorer and historian, describing lands close to the Illinois River in 1721

You need to know exactly where you are going. Either that or notice the phalanxes of corn marching by your car window have broken ranks. If someone has clued you in, told you a tallgrass secret, you'll be on the lookout for an unassuming sign amid the cornscape. The sign flags a tiny public parking lot for Nachusa Grasslands. Amble away from the road, away from the cornscape, past the sign—a few dozen feet down the trail, a world thousands of years old awaits.

I drive to Nachusa with bison, not corn, on my mind. These iconic prairie mammals, once flowing like roiling muddy rivers across

America's interior, would soon be welcomed back to Illinois as wild residents. I picture bison snorting blasts of steamy breath into the chill air. Bison hooves pawing the chocolate soil. Bison teeth ripping and grinding the golden grass. Bison barrel-rolling their one-ton bodies on this prairie patch for the first time in nearly two centuries.

John Schmadeke is waiting for me, having volunteered to be my prairie docent. Bespectacled, clean shaven, and a little on the serious side, at first glance John appears more a data analyst for an IT company than a nature lover. A bit of a prairie trick to that—like the grasslands that surround us, John contains dynamism under the surface.

I ride shotgun as John drives us toward the interior of Nachusa, giving me the 3,500-acre lay of the land. Sitting in a pickup feels surprisingly good, seat belt holstered, bouncing down a narrow road, grasses whispering against the doors and tickling the truck's underbelly. The terrain triggers memories of teenage fishing trips in Oklahoma with my buddy Cory Woodham, washboarding over land his grandpa leased—once a place of adventure, now a suburban golf course, more but not better used. The windows are down, the smell of sun and soil and sky blows back my hair, and the freedom of stopping where we please, going when we want, envelops me.

John catches me up on Nachusa's history as my eyes relax, my unfocused gaze absorbing the sun and shadow entangled within the wind's undulations. A hazy sense of Nachusa's past gains clarity. The pickup rattles us through space and time: we are navigating a fortunate remnant, a slice of land that escaped a fate that befell so much of the nation's tallgrass. Strip malls, suburbs, and agriculture account for the disappearance. Shop on it, pave over it, or draw our bread from it, but prairie left to pursue its own wild ends is an Illinois anomaly.

Nachusa stands as an exception to the rule of corn and soy. Beginning with 125 acres purchased by the Nature Conservancy (TNC) in 1986, this prairie parcel slowly expanded as TNC acquired slivers of neighboring farmland and restored their wildness. Now Nachusa can be described as a botanical hot spot, life support for almost 750 native plant species. It is also a refuge to animals like the Blanding's turtle, an endangered creature whose yellow-stippled, inky shell resembles a primitive nautical star chart. This chart, paired with an endearing smile, leaves the impression that the turtle knows where he's navigating, even when we humans don't.

In terms of biodiversity, John remarks, Nachusa is "like the Florida Keys or Hawaii, but with much less attention given to it." The flashy neons of plants and animals in those tropical locales may induce spontaneous *oohs* and *aahs*, but I am quickly learning how prairie offers more subtle varieties of spectacular. No misty mountains or tree cathedrals—for prairie, a person must get closer to the ground. Hands and knees might be best.

John stops the truck and we clamber out at the base of a small rise. A gentle walk up trail brings us to an overlook, which John confides is a favorite spot. It is easy to see why. From the proper angle and with modest imagination, the Illinois past comes into view—a rolling, sun-flecked prairie punctuated with gnarled bur oaks, aflame in copper and amber grass, stretching and yawning into the distance.

Little bluestem dominates here, responsible for the ruddy palette at eye-level. From breaths of wind, this grass decants various shades of whisky into its sturdy stalks. In the midst of these copper stems, other glories abound—from tiny four- and five-petaled lavenders a few inches from the ground to soaring talons curled beneath red-tailed hawks above—all of it anchored and nurtured by a soil beholden to time scales infrequently pondered.

Farther into the tallgrass, our next stop-and-walk delivers us to a handful of prairie "knobs." Nachusa lies within a fifty-mile oval of sand and gravel, folded together like a lumpy blanket. The knobs' significance, ironically, derives from their nutrient-poor soil. While neighboring fields tasted the blade of the plow as they were rutted and converted to farmland by industrious settlers, the thin soil of these knobs made them seem unworthy of the effort. As a consequence, many of the knobs host significant rare plant communities, like reliquaries laid out in the open sun.

On one of these knobs, John points out the pale purple sunburst of a stiff-leaved aster and tells me about a few rare species for whom

Nachusa is critical, such as the regal fritillary, a butterfly rarely found east of the Mississippi River. Cinnamon orange with creamy dabs of white on its hind wings, the striking two-inch insect lays eggs on bird's-foot violet, a plant that needs the open sun that a knobby habitat kindly provides. I poke around the knob, tracing my thumb along the stem of an aster, thinking about relationships between plant incubators and insect pollinators and the way a violet can beckon a butterfly. We next descend to the calcareous soils of a fen, a groundwater-fed, mineral-rich depression that collects water and hosts a unique assembly of creatures. Because it is October, we've missed the profusion of summer colors, but John thinks we might at least find the last iridescent lavender-blue blooms of a fringed gentian. We do. If someone had shown me flowers like the bird's-foot violet or fringed gentian early enough in life, I would have become a botanist. Or if I had the choice, a regal fritillary.

As the day wears on and John escorts me farther into the prairie, gratitude occasionally swells my chest, along with the bittersweet tang of knowing that less than one-tenth of a percent of Illinois prairie has survived the plow and the pavement. Places such as Nachusa open portals into the historical past and glimpses into possible futures—but sometimes it's best to take a breath and appreciate the gift of this living world in the present.

Nowadays prairie doesn't burst forth on its own. Interpersonal or interspecies, all relationships demand work, and healthy prairies require commitment. John tells me that invasive plants—like red clover, parsnip, and honeysuckle—present a perpetual problem. These nonnative species can quickly colonize a patch of soil.

John knows this firsthand. A steward at Nachusa for over two decades, he understands that the prairie depends on constant maintenance and volunteer gumption. To put some numbers on this, a prairie planting from scratch uses, per acre: forty-five pounds of forb seed (flowering plants) and five pounds of grass seed, and 350 hours of work per acre, over a three-year span. That's just to get things going. Volunteers log many additional hours monitoring their assigned sites; weeding invasive plants; and collecting, cultivating, and sowing thousands of tiny native-plant seeds. Most of the cultivation is done by folks who live nearby, though some people make the pilgrimage from farther afield in order to throw their backs into the collective effort.

Cultivation suggests domestication, diverting a plant away from its wild origins, favoring certain characteristics that favor us. Cultivating prairie thus brims with ironic and resurrective possibility. Each shovelful of soil turning the earth from farmland monoculture into vivacious, diverse polyculture also turns the trajectory of the land: from tame *thing* to wild *being*.

These acts of prairie cultivation strike me as radical, literally reaching to the roots of our relationship to the land. It was once believed that prairie needed to be brought under control, civilized by the farmer's hand. Sewell Newhouse, who designed a steel trap in the late 1800s that was used widely by Western forest rangers to capture wolves, provided a summation of this cultural perspective when he commented that such traps were "the prow with which iron clad civilization is pushing back barbaric solitude, and is replacing the wolf with the wheat field, the library, and the piano."[1] Like the wolves in Illinois, the prairie's "barbaric solitude" threatened the pioneer's sense of good order, which amounted to a land in the firm grip of human domination. Cultivating prairie, in contrast, unearths the will of the land. The power of prairie restoration derives not from denaturing in favor of so-called civilization but in unleashing care for the land's will, an active acknowledgment that there may be desires greater than our own that spring from the soil.

John and I make an impromptu stop along the way to greet Jay, a longtime volunteer known for developing successful protocols for newbie stewards, easing their transition into caring for prairie. After some shouting, we find Jay chest-deep in the grass. I shake his hand, noticing his soiled nails and knuckles, earth darkened from collecting seeds. Our brief conversation, touching on hard living and circling back to the pull of prairie, makes it clear to me that Jay knows this place, that part of him has come to reside here. I expect that for many stewards, Nachusa has become a cathedral without walls, a bodily connection forged with sweat as well as a refuge for the spirit.

How do the bison figure into all this? For several years, part of the stewards' labors at Nachusa has been to make this prairie as hospitable as possible for bison. Their housewarming gifts are diverse prairie grasses.

The bison are expected to do their part. Prairie, to continue to be prairie, relies on some essentials. A person can't merely plant native seeds and hope for the best. Restoration stewards have learned, for example, that fire is critical—for beating back encroaching trees, suppressing nonnative species, and cycling nutrients into the soil for fire-adapted native plants. But fire is only one part of the equation. Prairies were grazed by bison, and bison preferred certain plants for their foraging. The shaggy wanderers sculpted the land with their migrations and nutrient cycling (i.e., eating and pooping). They also created ephemeral pools of water because of their famous penchant for wallowing—more likely in happiness than in misery—tossing their bodies back and forth in the dust. The upshot is that bison habits diversified habitats, creating new mosaics for various expressions of prairie life.

Before the bison arrived at Nachusa, and despite a regular regime of controlled burns to emulate fire disturbance, the plants seemed to lack a key ingredient to their flourishing. Some native grasses were dominating; others, despite repeated seedings, couldn't find a roothold. A big part of the problem was the absence of grassland grazers. When discussion of introducing large herbivores began at Nachusa, cattle briefly received consideration as a surrogate for bison. After the dust of deliberation settled, key Nature Conservancy leaders and stewards determined that nothing could replace the ecological impacts and benefits of bison, the prairie's original grazers.

Toward the end of our tour, John takes me to a place where the bison would soon reside. It's a spot at Nachusa he knows well. He and his wife, Cindy, were stewards for six years of this parcel, the Hook Larson Unit, a 140-acre tract that would become the initial location for the bison reintroduction. I graze the palm of my hand over the fuzzy tops of little bluestem by the roadside. John gestures at the ubiquitous clumps of these grasses. Little bluestem is a bison favorite and they will eat it right to the ground, he tells me. The result will be openings for a variety of sedges, grasses, and flowers. Many animals stand to benefit as well, such as the upland sandpiper, whose cereal-bowl nests scraped out of the soil depend on large patches of low-lying vegetation. Sandpipers are usually shorebirds, but this sandpiper roams between seas of grass. Bison create the necessary islands.

Thirty bison from South Dakota's Wind Cave National Park will soon be transferred to the north side of Nachusa. The plan allows for the herd to expand to one hundred animals, eventually to roam 1,500 acres of land.

"After that?" I ask John.

"We'll just see," he replies.

Before getting back into John's truck, I pause and scan the rolling landscape of Nachusa and the nearby farms. John tells me about other prairie patches in the area, tangles of wildness he would like to see linked. He envisions a future prairie pathway across the state. For a moment I don't see land. I see water. In my mind's eye, I see islands of tallgrass prairie, a diverse tangle of life separated dot by dot like an archipelago. These islands reach for one another through the corn, longing to be joined and reconnected. Between them, dormant seeds wait to be loosed, perhaps by the hooves of bison, in a consummation of life.

．．．

If we could ask bison what kind of prairie reunion celebration they'd prefer, I wonder what they'd choose. I suppose bison wouldn't care so much about the ceremony as the fact that they were back together again. Probably a good thing, I think, as Smokey, representative of the US Forest Service, rests a friendly paw on the back of a woman who is holding the purple handles of an oversized pair of scissors. The dignitary with the scissors leans forward and snips a giant ribbon outstretched before her. Applause, along with a few whoops, bursts forth from those of us forming a semicircle of onlookers. Smokey claps too, then waves with both paws at the crowd. No one appears the least bit fazed. Perhaps a shirtless, smiling bear in blue jeans is no less surreal than what we are here to celebrate: the return of wild bison to yet another Illinois prairie.

I am an hour's drive southwest of Chicago—a shockingly small distance from one of the largest cities in the United States—at Midewin National Tallgrass Prairie, almost twenty thousand contiguous acres of land under various phases of restoration. By the time I watched Smokey give his blessing at Midewin, the bison at Nachusa had been confidently roaming the land for a little over two years, blending in as though they'd never been absent. One bison herd on one prairie is fantastic. Two bison herds on two different prairies begins to feel like a movement.

A few hundred yards from the crowd here at Midewin, a cluster of bison moseys along a fence line. I lock eyes with a formidable-looking bull who does not, it seems to me, particularly care for the fact that I'm admiring him. I wish fences weren't necessary and that bison could roam anywhere they please at Midewin. For the moment, though, the barriers seem a supremely wise investment. I avert my gaze so as not to tempt fate.

Instead I watch two calves prance, skipping and bouncing their burnt-honey bodies between the legs of their kin. This is a new generation, the firstborn bison of Midewin, another milestone. These calves at play represent something new and simultaneously very old and wild, like seeds unearthed from an ancient lithic midden and resown in the present. Tender shoots of hope rising to gather sun.

The name of this prairie complex, *Midewin*, is worth dwelling on. The Potawatomi root of the word, *Mide'*, indicates something "mystically powerful," and *midewin* refers to the ritual curing societies that once played an important role in Ojibwe, Ottawa, Chippewa, and Potawatomi cultures.

The land was renamed Midewin once the US Forest Service took responsibility for it in 1996. A different name and a different purpose marked this spot a few decades ago: the Joliet Army Ammunition

Plant, a full-scale military operation responsible for manufacturing and storing thousands of tons of explosives intended as the blunt force for winning wars. A few hundred cavernous concrete bunkers still stand, their interiors ripe with cool spectral air. A place that once dealt in death has been reconstituted to serve life, and a sign at the Midewin visitor's center speaks to the reemergent spirit: "Where people and the prairie restore each other."

A friend of mine who is a steward at Midewin, Arthur Pearson, witnessed the initial bison release at Midewin. The thunderous reverberations of the bison's hooves, through the soles of his boots into his expanding chest, produced an experience he says he will remember for the rest of his life. Arthur is here today to wish the bison well.

After the ceremonials, the two of us say goodbye to Smokey and visit a portion of Midewin that offers a vivid visual summary of the loss and restoration of the prairie. Row upon row of various native prairie plants, in different phases of growth, extend for acres in parallel lines. Because of the orderly lines, the arrangement resembles the work of an aspiring farmer who got his addresses mixed up. Yet this is how prairie must be grown to meet the demand in a state—known as the Prairie State—that cannot muster enough prairie seeds on its own for large-scale restoration. There's just not enough naturally occurring seed left to go around.

While Arthur and I gaze at this odd arrangement of cultivated wildness, he ruminates on how Midewin will be bursting with color in a few months once the summer sun raises the grass to its full height and the blooms begin to unfold in crimson, cream, and indigo. "This place is so special," he says wistfully. His voice breaks on the word *special*. He apologizes, pinches his nose, and gathers composure, adding, "It's just so beautiful to see the prairie come to life." The pooled tears in Arthur's eyes allow me to glimpse the many years, community meetings, early morning reveries, and late-evening muscle aches that created abundant life where once there were only instruments of wartime destruction.

Where people and the prairie restore each other.

A week later, I return to Midewin for a bird walk with Arthur. I'm a casual bird guy, catch as catch can. Arthur is much more experienced. A short distance down the trail, the differences between the two of us are immediately noticeable. I walk with my eyes; Arthur walks with his ears. I scan the grass tops for flashes of movement and variations of color, while Arthur frequently stops and tilts his head toward some queer trilling, barely audible above a blustery wind. "Bobolink." He smiles. Or, "Oh, wow, Henslow's sparrow." Or, "Hmm. Yep, indigo bunting." Then, he continues on his way, a satiated look upon his face, as I fumble with binoculars to link his sound with my sight. Unlike me, Arthur remains unperturbed when we can't locate a bird with our eyes. Later I realize that he is "seeing" the bird in his mind—his familiarity with matching species to sound creates a distinct mental image. It is as though he *has seen* the bird. Visual confirmation is unnecessary.

This came as a minor revelation to me, a person who relies so much on his eyes for sensory input. My perception felt truncated and one-dimensional in comparison to Arthur's. I was reminded that beauty can be delivered aurally. The world is tactile, and the human body is full of receptors. We are built to know the world with more than our eyes. I think of Thoreau, once astounded by the thundering of Walden Pond as the ice boomed in response to the sun's warmth, who commented, "The earth is alive and full of papillae."[2] We are sensitive beings reaching out to one another.

Where people and the prairie restore each other.

Because I live in the city, most of my nature time has an urban flavor. I champion small wonders—the spiky geometry of a sycamore seedpod, the droning buzz of cicadas, the jewelweed in the sidewalk crack. Walking in Midewin jostles loose something that urban living had shouldered to the back of mind: for many animals and plants, especially the rarest, size matters. There is more than awe and sublimity at stake. Many species need large blocks of habitat, whether for foraging, for mating, or for moving around. One expert whom I heard speak at a conference estimated that fifty thousand acres were necessary to restore ecosystem function with bison. It may be true that worlds of wonder are available in between sidewalk cracks, but most creatures cannot live on wonder alone. Maybe there's an aphorism here: let wonder begin in the cosmos of the backyard, but do not let it end there.

Arthur also lives in Chicago, but his draw to wild things led him to Midewin, deeper and deeper into what the prairie offers. When we walked together, we discussed how it might be possible to get people interested in the subtler beauties of the prairie. Tallgrass isn't a landscape to which everyone feels immediately drawn. A peaceful pine forest, sure; a sun-dipped blue lagoon, of course; a fierce granite mountain, yes. Prairie requires attention to love, but once one is familiar with the worlds it contains, it's hard to see a prairie savanna as anything less than precious.

In one of Aldo Leopold's most beautiful essays, "Marshland Elegy," he reflects on the aesthetics of a marsh, a place that would not be many people's first pick for a beauty contest, even the environmentally inclined. "A sense of time lies thick and heavy on such a place," Leopold observes. As for the marsh, so for prairie, which is really a less-wet marsh. Leopold looks at the marsh and sees not merely what lies before his eyes; he sees millennia: "An endless caravan of generations has built of its own bones this bridge into the future, this habitat where the oncoming host again may live and breed and die." A fuller aesthetic emerges by contemplating the vast sweep of geologic time, cycles within cycles of life and death and rebirth, laid down in layers and embodied in the creatures who now are able to inhabit and thrive in such places. "Our ability to perceive quality in nature begins, as in art, with the pretty," Leopold writes. "It expands through successive stages of the beautiful to values as yet uncaptured by language." This, he surmises, is natural beauty of a "higher gamut, as yet beyond the reach of words."[3] We can't wrap human language around such beauty, but we can feel it somewhere deep within.

Flashy baubles initially capture a person's attention, affections, and fascination. But the maze goes deeper. Bison are a great example of the way an animal can focus people's care. They've sparked the curiosity of many who wouldn't otherwise bother with Midewin to wonder what exactly is going on. Bison have, in effect, become ambassadors of prairie, living history lessons, portals into what makes a place wild. We begin with the pretty. Or the imposing. From there values can expand into that higher gamut of which Leopold spoke. You may start with bison, a good place to start, and they may lead you down a path that ends up changing everything you see, everything you hear. The world becomes renewed, like the prairie in summertime or after a burn. Sometimes you must burn down what you know to get something new to grow. With a cleansing fire, light can reach the seeds that were always there, patiently waiting under the honeysuckle brambles for the welcome face of the sun, waiting to produce a beauty beyond the reach of words.

Where people and the prairie restore each other.

Odd as it may seem, bison have helped me better understand the city. What happens in the city doesn't stay in the city. Because of its rapid growth, Chicago in its formative years was often described with metaphors that evoked the youthful, muscular body of an adolescent. Michelangelo's *David* gazing out upon a prairie Goliath that was destined to be laid low. My hope is that a greater wisdom can be found beyond those teenage years of bravado. Bison may have a role to play. Recently highway plans were scrapped, in part because of the predicted environmental impacts that would have resulted from a lopped-off corner of Midewin. Still, billions of dollars continue to pour into transportation corridors that spiderweb the landscape with impermeable pavement, fragmenting what remains of the fragments that other species rely on for their very existence. Could we reach a turning point and decide to spend equivalent energy and finances on opening pathways for a living landscape? Large mammals like bison, because of their need for large habitats and their key roles in shaping those habitats, may point the way toward alternative kinds of transportation corridors—for the four-legged instead of the four-wheeled.

An evocative term describes the improvisational pathways that pedestrians create as they navigate between buildings and streets:

desire lines. These non-sanctioned trails are everywhere to be found. I've often added my own affirming footfall to their existence as I cut across the corner of a courtyard or stray from the sidewalk in favor of a more sensible, less constricted route. *Go this way*, the pavement says, and people respond, *No, thank you*, opting for another way. Desire lines emerge most conspicuously after a snowfall, especially when the sidewalks and streets are freshly blanketed. Geometry intended to direct foot traffic cedes to intuition and spontaneity, and the impromptu pathways bear silent witness to the evidence that while city engineers tend to think in grids, most of us do not.

I first came across desire lines while reading about Detroit's travails following the constriction and collapse of the city's industries. High rates of vacancy are now common throughout the city, with some blocks almost completely emptied of residents. In the midst of a fading former glory built largely upon the automobile, people began to ask, *What now?* There might be an opportunity, many people think, to construct more durable forms of existence, to follow the contours of the land with more fidelity, to rely upon the local community's means of production and subsistence rather than the volatility of markets in other lands. As urban agriculture spreads, sustainable economies get a hearing, and local decision making gains priority, a remarkable knit of desire lines has sprung up between Detroit's ruins, cutting across open fields, weaving between neighborhoods, and revealing foot-friendly modes of transport.

While it remains to be seen what shape a Detroit of the future will take, seeing images of these desire lines prompted me to think about other lines closer to home—not just about the footpaths that humans create when they are not hemmed in but about the nonhuman desire lines that exist all around us. What if we favored the living with fewer fences and instead fenced our highways and roads, saving ourselves from beasts of our own creation? What if we gave deference to the desire lines that other animals create? What would the Prairie State look like if we followed the desire lines of bison?

My hunch is that if we attend to their desires, continue to restore the grasses that provide their sustenance, allow them to reshape our lands and our vision for what these lands can be, something very special indeed will come into being: a people who know how to live within a world of wild values as yet uncaptured by language.

．．．

I say goodbye to Arthur and pull back onto the two-lane pavement. Manicured rows of corn flash by. I am accustomed to hearing environmental issues talked about as rearguard actions: how much we've lost, what we are now losing, what can be salvaged. Large-scale restoration projects such as those at Nachusa and Midewin flip that script. They beckon our imaginations toward envisioning what can be done, how much can be restored, what species we might welcome home.

What do prairies need? It's a question worth asking again and again. The prairie's needs and ours are mutual. The answer lies near Leopold's notion of beauty in a higher gamut, something difficult to articulate, maybe impossible, but felt as a healing, a reknitting of our own lives into the larger story of place.

Landscapes contain visible archives of a society's values. The bison at Nachusa and Midewin offer a hopeful nod toward alternative values and pathways. These lines of desire reach from past to present, spreading themselves from root to sky. Even the smallest step in response can bring any person closer to the Midewin creed.

Where people and the prairie restore each other.

Corridors of Change

Greenways

Blueways

Mindways

Greenways

These pedestrian movements thread a tangled mesh of person-
alized trails through the landscape itself. Through walking, in
short, landscapes are woven into life, and lives are woven into
the landscape, in a process that is continuous and never-ending.

—TIM INGOLD, *Being Alive: Essays on Movement, Knowledge
and Description*[1]

Orienting

The driver fastens a white-knuckled grip on the steering wheel and
aims the car my direction. Eyes locked on the road, he punches the
horn with a fist. I draw a breath, waiting. He shoves the pedal down.
The car accelerates . . . passing directly under my feet.

I am standing on the 606 Trail, an abandoned elevated train line
recently converted into a multiuse path. Beneath me is Western
Avenue, a busy urban road rife with traffic, treacherous to the city
pedestrian. But I'm secure on my perch, watching eager drivers determined to reach the next red stoplight. I raise my head and glance

down the trail, my own sense of urgency compelling me forward in order to witness everything a walk above the city has to offer.

The 606 Trail, so named for the first three zip code numerals shared by Chicagoans, is a journey through a variety of well-planned surprises. About ten years of community meetings and public-private partnerships fed the final incarnation. Early design mock-ups and planting proposals aimed for aesthetic impact, with trail sections receiving dreamy monikers such as "Evergreen Spires," "Blue Meadow," "Urban Savanna," "Echinacea Field," and "Sumac Tunnel."

The bur oak sentinels and paper birch trees, head high at present, will inevitably stretch their fingers, topping the curving lamps along-side the pathway. Tufts of prairie dropseed grass—resembling the hairdos of Dr. Seuss's Thing 1 and Thing 2 in *The Cat in the Hat*—will eventually cover the bare patches of exposed ground. But already this newborn urban corridor, 2.7 miles long, sixteen feet above the ground, and an average of thirty feet in width, presents a living palette throughout the seasons.

Color, structure, and seasonality—even phenology, the timing of flower blooms—are all accounted for in the trail's linear east-west design. A striking example is the serviceberry trees (*Amelanchier grandiflora*), whose ivory-pink buds express a sensitivity to minor temperature differences. At strategic points along the corridor, over a five-day period, the staggered timing of serviceberry blooms announces the coming spring and displays the cooling effect of Lake Michigan. One writer likens the blossoming to "lights turning on down a long hallway as the city slowly comes back to life."[2] The trees embody a teaching moment about the city's interwoven habitats, a way to understand and measure shifts in microclimates by attending to beauty as it unfurls.

Whenever I walk the 606, I'm caught up in its concentrated energy. On a busy weekend, the central concrete trail feels narrower than it is. As often as I can, I take one of the alternative, pedestrian-only footpaths, tendrils of crushed gravel adjacent to the main trail that wind through miniature forests of birch and evergreen. I'm not quite shielded from the cyclists' blurs but feel grateful for the reprieve.

The elevated path summons people like a green magnet. For some, it functions as an alternative exercise track, a way to squeeze a jog or bike ride into the workday. For parents pushing strollers or clasping toddlers' hands, the trail offers safe haven where they can forgo the standard parenting protocol of looking both ways at every intersection. Also rambling about may be a handful of people like me, who relish viewing city streets from this vantage, who feel in the soles of their feet a vibrant urban future.

At Milwaukee Avenue, all three major streams of transportation intersect—foot, car, train—with trailgoers passing between the street below and the elevated tracks slicing diagonally above. The city's layers can be exquisitely felt here: the "L" train clanking through the canopy, pedestrians and cyclists sharing the understory with the rooftops of three-floor walk-ups, and automobiles possessing the forest floor. Layers of the urban biome.

In the late nineteenth and early twentieth centuries, freight trains in Chicago ran at street level. A troubling number of injuries and deaths prompted city officials to mandate that the trains be raised, which produced the track on which I now am walking. A transposing of transportation: the industrial trains are gone and the pedestrians have been raised to avoid the cars.

Rail-to-trail conversions possess a special allure. The by-products of trains—noise, pollution, and physical barriers—once dividing communities become corridors of connection. This makes the 606 feel like a broken bone on the mend. The trail transects four neighborhoods—Logan Square, Humboldt Park, Bucktown, and Wicker Park—that a few decades ago epitomized the rough-and-tumble, working-class grit of Chicago. Midcentury novelist Nelson Algren, who lived in Wicker Park for over a decade and chronicled the lives of the city's outcasts and down-on-their-luck dreamers, memorably described this version of Chicago in his essay "Chicago: City on the Make": "You may well find lovelier lovelies, but never a lovely so real."

But the city is an organism of change. By the mid-1990s the small manufacturing enterprises and textile factories that would have been familiar sights to Algren were long gone and the train traffic had ceased. Any visible signs to show this portion of the city's links to industry remained only in its lingering lack of greenspace. One of the neighborhoods that intersects with the trail, Logan Square, possessed the least amount of open public space in the entire city. This lack seeded the desire for the 606.

The 606's rail bed is mostly ramrod straight, yet the trail curves pleasantly, both side to side and with gentle vertical rises and dips, like a slow undulation of sound shaken across the city. Walk the full length and you'll experience an aural gradient that seems to ride this wave. The eastern trailhead rubs close to a continuous thrum of cars on Interstate 90/94. Construction saws whine and hydraulic lifts groan, signaling warehouse conversions to posh loft spaces and high-end condos. Sounds become more subdued as you go farther west, away from the highway and the Metra tracks. Distant train horns, the puttering of idling trucks, and scattered car honks cede air to gossiping sparrows and the clicking burbles of starlings. The *tap-tap* of car horns is replaced by the *tap-tap-tap* of downy woodpeckers. Two-flats give way to single-family homes with postage-stamp yards, playground laughter from a few nearby schools, and some still-vacant lots. Juncos flit between the purple coneflower, little bluestem, and silvery sage, luring pedestrians to the end of the wave.

The western terminus of the 606 diverges. A walker may descend to street level and return to the workaday world, either going north or south, back to the grind, back to the street traffic. Or the walker can go up, accepting an invitation from an ascending spiral pathway. This latter path orbits grassy slopes until the gravel opens upon an oblong-shaped clearing. Known as "the observatory," this end of the trail provides a modest finale, a crow's nest for watching a blood-orange sun set over the city or a raised dais for contemplating the stars. The slight elevation creates seemingly infinite sightlines across the horizontals of Chicago.

This trail feature is loosely modeled on an ancient solar observatory in Chankillo, Peru. At the western end of the oblong clearing, a low-lying cement wall juts from a grassy slope that doubles as a place to rest a tired back. Three notches in this retaining wall divide the year, framing the points on the horizon where the sun sets during solstices and equinoxes. A cement *X* crosses the ground near the center of the opening, pointing at two of the notches, marking the path of the sun at its winter and summer extremes. Just outside the oblong clearing, a simple sculpture—looking something like an oversize dial for a manual timer with the word *equinox* chiseled on its side—points toward the middle notch.

I slowly pace the oval circumference, hands clasped in the small of my back. I stop at the squared notches in the retaining wall, bending to look through each one as though it were a spyglass for viewing the western spread of the city. I raise up and arch my body, chest facing the sky. Somehow the strangeness of the planet tipping toward and away from the sun feels more real. Like I should be grateful I don't shake loose. At the same time, I feel more bound to everything: we're all sailing through space together around that big star—the bur oaks, the downy woodpeckers, the purple coneflowers.

When I first began working in downtown Chicago, other than all the revolving doors, one of the architectural oddities that stood out to me was the innumerable clocks in the city. At some street intersections, multiple buildings thrust giant timepieces over the sidewalks. This decorative infatuation with clocks quaintly testifies to lifestyles at the turn of the twentieth century, when not everyone was carrying an iPhone, but it also advertises a particular philosophy of place. This city prides itself on a can-do work ethic. Time is money. Those clocks remind passersby, *Aren't you supposed to be somewhere doing something? Hurry along now. Chop, chop.* From its pioneer origins to its industrial heyday, Chicago has always been, to quote from Algren again, "a city on the make," in the business of doing business. It was built for it. Chicago's railroads opened the western frontier. White pine from the North Woods. Cattle from Texas. Grains from the fruited plains of Nebraska and Kansas. Chicago was the linchpin that linked distant markets, collapsed space and time, and made millionaires out of merchants, steel men, and land speculators.

The earth and stone sculpture of the 606's observatory gains greater meaning in this context of busyness. The equinox and solstice notches mark a deliberate redirection of attention: to the deeper cycles that all our daily artificial microdivisions of time rest upon. These larger orbits touch your skin as the bite of winter, lure the flowers out of hiding in spring, send the butterflies skipping across the grass in summer, and ignite the crimson fireworks of sumac trees in the fall. The observatory reminds me that our paths are entangled with the lives of other beings, all of us dependent on that sun now going down on the western edge of the city. It presents a place of quiet presence, inviting sun-hungry bodies out of the enclosures of cars, off of the streets, away from the clocks. Entering or exiting the trail, a moment of pause: a reminder of what a city and all its creatures depend on for sustenance. A grateful look skyward for what truly keeps all the layers of the urban biome humming.

Wayfaring

Why do I walk? I press my toes against the edges of that question and realize I haven't always been a walker. I was raised in the car-almighty suburbs, where a person needs to get behind the wheel to get anything. Long-distance jaunts on foot were special occasions, reserved for visits to national parks and scenic lakesides. Perhaps counterintuitively, only when I moved to the city did walking become an imperative—a primary mode of transport, contemplation, and enjoyment.

In the city, my family can rely on a solid network of public transportation, so we found we no longer needed a car. The relatively short distances between essentials—grocery store for our food, school for our son, coffee shop for our fuel—made traveling by foot sensible. But walking is about more than the essentials. Walking unearths the nooks and crannies of the city, affirms the city as more than its pavement, reveals a city as a living place. Perhaps that is an essential, too.

"It is as wayfarers," says pro-bipedalist and anthropologist Tim Ingold, "that humans inhabit the earth."[3] By which he means, our embodied experiences of moving along pathways is how we know a place. As he puts it, we know *by way of*—on the path, around the bend, over the river and through the woods. In cities shaped by the authority of the automobile, most urban dwellers are likely accustomed to thinking of places in modular terms—each destination an IKEA furniture store, a collection of distinct *things* to which we travel.[4] Spots on a map, widgets in a box, nodes in a roaded network. In contrast, Ingold suggests we live within a *meshwork*, each of us a knotty entanglement of relationships, a squiggly intersection of pathways of experience.

In the meshwork of wayfaring beings, a place acts as a verb, not a noun, generated by lives in motion. The wayfarer's movements are not place bound (creatures in the container of the city) but place binding (creatures whose movements are the ongoing creation of this thing we call the city). Fellow wayfarers, we thread our way *through* the landscape not *across* it or *over* its surface. Indeed, we are part of what creates the surface, tracks in the snow, traces in the wind. Stitching our lives through the landscape, we generate—bring into being—the textures and the stories of a place.[5]

I've done a lot of wayfaring in the city, hoboing it beside rail tracks, secreting my way through alleys, sucking in lakeshore breezes with feet dug into sand. Mornings I've watched the sun rise and pluck itself from the lake's grasp; evenings I've flung my thought arrows at Orion's belt. Winters I've plowed my legs through snow drifts in subzero temperatures, jewels of ice dangling from the edge of my toboggan cap; summers I've spread my toes against cool grasses and punted dandelion seeds into the wind. Many miles, overlaps and tangled knots, tying together countless tracks, the warp and weft of footpaths, sidewalks, desire lines, downtown streets, and forest trails.

I'm not the first to value walking through a city as a way to know its inhabitants more deeply, including nonhuman animals whose ways of life can only be approached on foot.[6] I have done a small bit to spread this gospel of urban wayfaring. Countering a tendency toward introversion, I recently began organizing an urban nature-walking group with my friend and fellow peripatetic Liam Heneghan.

Liam makes for an excellent walking companion. He's an Irishman with a fondness for spinning a yarn. The man also doesn't lack for opinions, typically darkly humorous and observant ones. A soil scientist by training but a poet by instinct, Liam looks the part, bespectacled and brandishing a long salt-and-pepper Whitmanesque beard that serves as a stray pencil holder. Before we formed the walking group, Liam and I got to know each other by tramping through Chicago together. On these rambles, we grouse about writing and discuss current loves and loves lost, and I frequently hear about his latest schemes—ranging from personal tests of will, such as subsisting on a diet comprising solely of astronaut food, to more formal academic endeavors, such as research on the decomposition of feces. From him, I've picked up Irish double negatives such as "that's not nothing" as well as an appreciation of the tin whistle, an accoutrement always tucked into his vest pocket in case inspiration strikes.

Perhaps Liam first suggested that we spend some time in cemeteries. He's a bit of an arborist, so this would make sense, since many cemeteries serve as accidental arboretums, protected areas where trees from around the world can be found. Besides parks and some city shorelines—though much less frequently trafficked—cemeteries most closely approximate an intergenerational public commons. Truly intergenerational. Seen from above, they become islands of green surrounded by the dark waters of pavement, clearly set apart and honored as sacred spaces. Few natural areas in the city garner that degree of deference. The workaday world is not allowed to intrude; convenience stores shall not pass.

Cemeteries bend time, revealing wayfaring as more than a present-tense activity. This type of path making includes those who have come before us—our traces reach through place *and* time. When walking among the gravestones, you walk lands that respect the dearly departed and, by default, provide habitat to the still thriving.

Graceland Cemetery. A glorious autumn day. Maple trees in spasms of gold and scarlet. Liam chose this cemetery for our group walk since it's one he knows well. About twenty of us are here, eager to see the interior of one of the larger patches of city greenspace. Liam leads us along, stopping at various trees and adopting a role as charismatic undertaker. He talks decomposition, twirling an oak leaf above his head, drawing the group's attention to the tannins that become visible in fall leaves as they go to ground, a tree's chemical defenses against herbivory now exposed in the rash of colors all around us. Beautiful poisons.

I pick up a maple leaf the color of a tequila sunrise, a dying ember descended from the sky. When I flip the leaf over, an intricate network of capillaries become distinct. The shape of the five-pointed leaf and its correspondence with the veins in my five-fingered hand is startlingly clear. A strong sense of kinship floods my body. Leaves are trees reaching out, the expression of their longing for the sun made visible in the imprint of these lifelines. We are all yearning for connection—the leaves above my head, beginning to turn toward their deserved seasonal rest; the roots underneath my feet, making slow traces through the soil in search of ongoing nurturance. For a moment, I feel I am merely an uprooted tree, wending my way through this cemetery in search of connection.

Liam continues to pontificate on the glories of leaf litter, while I look at the weathered headstones encircling us, thinking of other kinds of decomposition. There's something democratic about the work of soil. Graceland boasts a good number of elaborate obelisks, filigreed mausoleums, and unusual sculptural flourishes—hooded figures, Marian tributes, and angel wings—but time has taken a toll. Names and dates are fading; some of the older headstones offer only faint incisions in stone. Rain streaks and runnels continue the work of water, returning the seemingly permanent back to earth. No matter what fame a person accrues in a lifetime, no matter the monuments built to preserve that memory, cemeteries insist on recalling the larger cycles of which we are a part. Death generating life.

No matter who you are. The remains of a few notable Chicagoans now feed the grass at Graceland. The plot for Daniel Burnham and family is difficult to miss. He has his own island. Our group crosses an isthmus created by a wooden bridge onto a small, tree-sheltered grove in the midst of a lagoon. Here lies Burnham (1846–1912), Chicago architect and the brains behind the "White City" of the 1893 World's Fair. An intrepid visionary and Progressive Era champion of rational order and beautification, Burnham often receives credit for the phrase "make no little plans." Photographs of Burnham show a model, turn-of-the-twentieth-century cosmopolitan, a man of gravitas and well-parted hair, with a bit of a paunch and a bristled mustache that would provoke begrudging approval from a modern-day hipster. Burnham is best known not for his mustache but for coordinating the 1909 *Plan of Chicago*, a lushly illustrated regional masterwork that imagined Chicago's future in attracting (and keeping) wealth while satisfying the social and recreational needs of its citizens.[7]

This prime piece of cemetery real estate indicates Burnham's importance in the shaping of Chicago. While our cohort of walkers basks in the dappled sunlight of the island sanctum, a member of our group, Dan, nudges me with an elbow and nods his head toward his iPhone. The screen displays a snapshot of the grave markers on the opposite bank. Dan points with an index finger and I lean in for a closer look. Coyote.

I raise my eyebrows and look up at Dan. With his line of sight, he indicates the place where he aimed the camera. Nothing. Of course coyote—the boundary crosser—would be here. And then not here. Observing, disappearing, a cemetery mischief maker and ghost in broad daylight. No one else in our group notices. When I tell Hannah, an anthropology PhD student at the University of Chicago, of the coyote in our midst, she mentions that she and a team of graduate students recently conducted ethnographic research at another Chicago cemetery. Chinese families visit in order to leave offerings at the graves of their loved ones, including food items, honoring their departed ancestors. The food doesn't last long. Hungry ghosts? Hannah suspects the coyotes she's seen darting between tombstones are supplementing their diet with these ancestral offerings. The food may not reach the intended destination, but it feels like a fair exchange to me.

My focus returns for the moment to Burnham and his plans for Chicago. Anticipating the growth of the city, Burnham was among early advocates who sought protection for an outer ring of forests as a buffer from the sprawling human core. In addition to his proposals for boulevards and parks within the city—many of which still exist (good for urban wayfarers)—he saw the wisdom in setting lands aside outside of city limits before the tide of development swallowed the greenery. This brings us to a curious historical intersection of cemeteries, violent disputes between wealthy industrialists and the working classes, and forest preserves.

A different cemetery, Forest Home, lies adjacent to preserved land on the western outskirts of Chicago. The burial grounds host a striking bronze sculpture: the Haymarket Martyrs' Monument. The cemetery, like the forest preserves, once was outside of city limits, which is the reason the memorial is there. On May 4, 1886, in Haymarket Square, immigrant laborers and organizers gathered to hear speeches and rally together in solidarity for making the eight-hour workday a reality. Near the end of the rally, the assembly thinning because of inclement weather, 170 or so police officers marched forward to disperse the small crowd. Then, a bomb. Seven police officers were killed (likely due to the gunfire from the officers; only one officer was actually mortally wounded by the bomb). A hasty trial led to the conviction of eight prominent Chicago union organizers and anarchists. Four men, some not even in the vicinity of the rally when the bomb exploded, were hung. Burial of the bodies proved controversial. Under political pressure from the city's media and businessmen, the mayor refused to allow the anarchists' bodies to desecrate the city itself. So adding insult to death, the wrongfully convicted and executed troublemakers were exiled from the city and buried where the Haymarket Martyrs' Monument now stands.

After the riots, well-moneyed men of commerce from the Commercial Club of Chicago took the opportunity to safeguard the flow of capital, which was believed to be threatened by "foreign" subversives and their pesky demands for humane working conditions. They bought and donated land thirty miles north of the city for the express purpose of hosting a garrison of federal troops close to Chicago in order to quell further working-class unrest. Just in case.

This would come in handy eight years later for a current resident at Graceland Cemetery, George Pullman (1831–1897). Pullman's grave is not far from Burnham's, a short walk down a winding path after our group departs the island. Both men were members of the Commercial

Club, as are other Graceland residents. Our group saunters toward Pullman's resting place. Unlike others who are gone but not forgotten in Graceland, there's a good chance that Pullman's bones might not be making their contribution to the soil. That story has to do with the previously mentioned federal troops.

Pullman's railroad utopia on the city's South Side relied on immigrant labor. And at first, things were pretty swell. But good business can always be better. Pullman began looking for ways to increase his profit margins, which meant cutting the pay of those in his employ while keeping rents the same. (Pullman shrewdly owned all the housing in which his employees lived.) In the spring of 1894, a mutiny began in earnest. The federal troops for whom the Commercial Club had the foresight to buy land were called in to break the strike and get people back to the factory. Thirty strikers were killed. Things weren't ever quite the same afterward. Pullman died of a heart attack in 1897. Fearing the desecration of his body, his family had him buried in a lead-lined coffin, within a pit encased in a foot and a half of concrete, topped with steel rails and another layer of concrete.

As we stare and take pictures of the Corinthian column atop Pullman's grave, the knowledge of what is under my feet—a layer of history the city is built upon—struck me as simultaneously tragic, poetic, and completely in keeping with the sorts of boundaries that the attempted control of others necessitates. Pullman is an extreme example, but as I walk the peaceful paths of Graceland, additional signs appear. Money is needed to chisel elaborate grave markers, a last flourish of extravagance and status. I see masonic insignia on many headstones, signaling secret societies and male bonding, as well as last names that favor particular ethnicities and social classes. The work of the soil may be democratic, but the work to keep like with like takes cash and influence. I derive some comfort in knowing that no matter how thick the concrete between George Pullman's body and the soil, the roots will eventually find their way inside.

These thoughts flit through my head as our group arrives at what, for me, is the sacred heart of Graceland, though it is tucked into the far southeastern corner of the cemetery. The landscape here is unmistakably distinct from the well-manicured and orderly grasses surrounding it. Purple prairie clover seed heads and yellow coneflower bob in the wind. A path disappears into a tangle of tallgrass prairie. Field sparrows flash between the stalks.

Sometimes what is sacred is salvaged. Several cemeteries in Chicago—whether by happenstance, neglect, or as token tribute to the place's original appearance—host prairie remnants that stand tall in these feral corners, all that remains of the pre-urban landscape.[8] They are memorials to tallgrass, more important than any individual's headstone—Burnham's, Pullman's, or those of the other residents at Graceland.

I let the group outdistance me, hanging back and taking my time. I wind through the fingers of tallgrass and inhale the loamy scent. Seed heads are bursting open. Milkweed pods are split in two, the white comet tails of their seeds fluttering before launch. Most of the cemetery honors the dead. Here the honor is to the living. And the living come to remember, so that memories don't slip too quickly. These patches embody worlds that preceded the city, places to touch the sacred past-and-present-and-possible-future with one's fingers.

Cemeteries are typically places associated with solemnity and grief. Not for me. I come to the commons of Graceland to remember to live better, sweeter, to support those cycles that endure beyond a human lifespan. That's not nothing, Liam might say.

Passages

The economic and racial differences on display within urban cemeteries—so evident once you tune in to their wraithlike presence—do not rest in peace. One can wish that they would decay softly with the dead, marking another era, enclosed behind those magnificent gates as a memorial to wrongheaded thinking—but these divides scrawl themselves across the land of the living.

Chicago has a reputation as a segregated city, a city of immigrant waves, a city of neighborhoods, a concentration of American divides: between social classes, between ethnic groups, between haves and have-nots. I suspect a lot of people don't know how intentional that segregation was in the past and how it persists in the present.

City wayfaring reveals short distances between contradictions and sacrifice zones, the grit and the grace rubbing cheeks on blocks separated by a street corner or some invisible line of demarcation. Signs of hope, scars of abuse. You can't speed around these divides when you're on foot. Neither can you close the chasms through the simple act of walking—but you can witness the divisions, and this can unearth questions about how our city can be different, how it might adapt to cycles that nurture our spirits and exceed our lifetimes.

I get a dose of these contrasts when I visit the yet-to-be-constructed New ERA Trail, another elevated pathway that, according to some, is supposed to be the South Side's counterpart to the North Side's 606.

I can't recall when I first heard about the New ERA, but the idea of an elevated trail on the South Side immediately struck me as a hopeful endeavor. ERA stands for "Englewood Remaking America," an indication of the ambitious vision of those behind the initial pitch. Although the two elevated trails constituted separate efforts—with the idea for the New ERA springing forth several years after the 606—for a time news about the two trails was often paired, like box scores for the Cubs and White Sox. I occasionally checked for fresh updates about both trails. As energy (and money) poured into taking the 606 from beautifully wrought plans to reality, updates about the New ERA Trail ebbed. The 606 opened to deserving fanfare on a sunny summer day, but as far as I could tell, not a shovelful of dirt had been moved on behalf of the New ERA.

As an urban rambler, I wanted to see the prospects of the New ERA Trail in person. I finally arranged a visit, meeting two companions who knew the area much better than I—John Ellis, an Englewood resident and New ERA advocate, and Elvia Rodriguez-Ochoa, who works for a prominent regional conservation group. Elvia's organization helped facilitate outreach efforts that resulted in a community vision plan for the trail.

The yet-to-be New ERA Trail is intended to occupy nearly two miles of former rail line that runs along a berm, except when the path's concrete and steel bridges cross over the streets below. We walk and I pepper John and Elvia with questions. All the while, I keep thinking about the esteemed Chicago journalist Studs Terkel's description of Chicago as Janus-faced. Janus, the two-faced Roman god of beginnings and endings, departure and entrance, past and future, and not incidentally passageways.

The trail, still rich with vegetation on this sun-freshened September day, feels like a rope in a tug-of-war between possible futures. An informal path winds down the middle of the rail bed. Swallowtail butterflies dance between feral daisies. A gentle breeze whispers through the incomplete tunnel formed by a leafy canopy of box elder trees. Shadows stretch across lime-green clover sprinkled among remnants of wooden train tracks. A northern flicker arcs through the air in front of us, a patch of white above the tail feathers blinking out behind tree trunks.

Below, however, boarded windows of abandoned homes stare blankly at the street—empty husks serving as reminders that Chicago's industrial heyday is over. Due to a steep decline in manufacturing as well as the growth of the suburbs, the city lost population for four straight decades, from 1950 to 1990. The Englewood neighborhood, once a thriving commercial center in Chicago, contained ninety-one thousand people in 1930. By 2010, it had about thirty-one thousand residents. The area's depopulation is clearly visible near the trail. A total of eighteen acres of vacant lots hug the rail line, like a piano that's missing two-thirds of its keys. No Trespassing signs hang from chain-link fences and porches. Large empty parking lots sprout cracked asphalt and windswept trash. Open stretches with views of the streets leave me feeling exposed—a stranger and, maybe because of the signs, a trespasser.

I swivel my eyes while keeping one ear tuned in to John. He tells me of another major factor that accounts for Englewood's woes: the area was "ground zero" for redlining, an all-too-common practice by business prospectors, mortgage lenders, and real estate appraisers of grading certain sections of the city according to their potential for investment. The color-coded maps—*green* meaning "best" and *red* meaning "hazardous" on opposite ends of the scale—were color coded in another sense.

Following the Great Depression, from 1935 to 1940, the Home Owners' Loan Corporation (HOLC) created such maps for more than 250 cities across the country. The HOLC consistently identified African American areas or those areas "threatened" by African American residency as red. Researchers recently released online versions of these archived materials, noting that these maps "helped set the rules for nearly a century of real estate practice. . . . Indeed, more than a half-century of research has shown housing to be for the twentieth century what slavery was to the antebellum period, namely the broad foundation of both American prosperity and racial inequality."[9] Simply put, at a time when black people were migrating to urban areas in search of opportunity, redlining became a federally sanctioned attempt to keep black communities or those with increasingly black populations in their place.

When I later look up one of these maps online, I find the section of Chicago for the proposed New ERA trail.

Red.

Here's an excerpt about the area from the HOLC, written in 1940:

Many foreigners, Irish and Italian predominant. Substantial negro development exists between Aberdeen & Loomis, 60th to 63rd, even tho restricted. A Property Owners' Organization is endeavoring to hold that portion between 59th and 64th, Green to Carpenter, against colored infiltration. . . . A Copernicus school recess looks as if 90% of the children are negro. About the only way to stimulate white purchases is to rehabilitate homes and make them attractive. . . . An area with an uncertain future; transportation fair; known as part of the old Englewood district or West Englewood. The better class (or "lace-curtain") Irish have moved further south.

John, Elvia, and I pause for a moment. A man approaches in the distance, walking along the trail, but he disappears down the embankment long before we make eye contact. "Looking at the maps and then out your window, it would seem the red-liners won and did so a long time ago," John says. "The *persistence* of effects caused by decisions made almost one hundred years ago is breathtaking." The past hadn't gone anywhere. I felt overly vigilant and sheepish for feeling overly vigilant.

Other dubious practices too easily multiply grievances when part of the city is deemed unsafe. John surmises that one of the reasons the city has been less than proactive about trail construction has less to do with putting the final land acquisition pieces in place than with concerns about environmental contamination. If the trail moves forward, the city can no longer turn a blind eye to toxic soil and poor air quality in certain parts of the neighborhood. John suffers from bronchitis, and he attributes this malady to diesel particulate matter from a nearby railroad intermodal. High rates of cancer and pediatric asthma are common among his neighbors. "It's one of the city's dark secrets," he tells me.

A 140-acre rail yard, owned by Norfolk Southern, is mere blocks away from the eastern terminus of the proposed trail. The yard sees daily traffic from about a dozen trains and twelve hundred semitrucks—diesel-powered vehicles that sometimes idle for long stretches of time as they wait to be loaded or unloaded. A deal with the City of Chicago led to Norfolk Southern acquiring property on another eighty-five acres adjacent to the southern portion of the rail yard. An additional eight hundred semitrucks, per day, are expected to use the expanded intermodal once construction is complete.

Elvia later drives us through the area of the expansion. There's not much left. The homes are gutted or gone. Community advocates, anticipating increasing health problems in an already-beleaguered area, sought and received legal advice, eventually brokering a deal with the City of Chicago and Norfolk Southern. The good news: the agreement commits Norfolk Southern to install pollution controls, clean engine technology, and diesel filters on their trucks and heavy equipment, in addition to providing $1 million for sustainability initiatives and another million for job training in Englewood. The elevated line for the proposed New ERA Trail, which was owned by the railroad, also figured as part of the exchange for the expansion. In a tragic twist, residents who felt forced to sell their homes or move from the area claimed by Norfolk Southern's expansion—some of whom claimed the railroad expansion was a "land grab"—may never enjoy the benefits of the new greenspace. Janus again.

About halfway through our walk, we pass a painful symbol of the community's hardships—a shuttered school known as Bontemps. This vacancy may be the worst along the trail. Not long ago, the city attempted to address the Chicago Public Schools' budgetary deficit by closing over fifty schools identified as underperforming or underenrolled. Bontemps was one of them. The building could have a future, Elvia remarks, as we stare at the lonely school. Discussion is ongoing about repurposing the facilities as a vocational training site, with a focus on environmental jobs. For now, though, the school silently radiates loss. Dandelions and rust crawl up a swing set. Bars for climbing, polished by countless hands, are now polished only by rain. The blue springs and yellow chains unravel. Few silences bear the quiet as heavily as those of an empty playground.

There is more riding on the New ERA Trail than one additional walking route through the city, I realize. Turn the Janus face, another portrait appears. Signs of hope line the pathway. The visioning document is a thing of beauty, pastel highlighted illustrations and bird's-eye mock-ups of tidy urban farms, art installations, and parks providing fuel for the imagination. The aim for the trail, the vision plan declares, is to create a "regional destination with a unique identity that includes urban agriculture/horticulture, green energy and development interpretation, festival spaces and public art," including anchor points adjacent to the trail that will serve as inviting community gateways. One of the anchor points already exists—the Wood Street Garden, a thriving urban farm that sells produce commercially and plays a highly valued role in the community as a job-training site. There are many who see potential in Englewood for the expansion of gardens and farms, including strong networks among grocery stores, community markets, restaurants, chefs, and a local culinary training center. The trail would be a backbone for this network, offering vital gathering spaces of social interaction.

Images of possible futures for the trail in my head, we turn around and begin our trek back. Something crackles through the leaves, out of sight, just ahead. Coyote? Wouldn't surprise me. Coyote revels being at the boundaries, in between, on the edge of new creations. He doesn't need to wait on amenities such as water fountains and festival plazas and wayfinding signs. He knows the way forward.

We reach the spot of the noise. Some robins flipping leaves in search of insects; nothing else.

Later, we stumble across unexpected human noises. A group of seven or so men in hard hats are tamping down dirt, wielding shovels, and firing up saws. Early stage construction. We pause and chat with the crew supervisor and discover this is a work crew from Greencorps Chicago—a city-funded program that specializes in landscape restoration throughout Cook County. They are carving out a gently sloping path, connecting the top of the railroad bed to the ground below. It is only an on-ramp, an access point for further construction, but it is also evidence that the New ERA Trail may yet amount to more than good intentions.

Despite the gorgeous day, I can't escape the feeling that I am walking through wounded lands. It isn't just the historical and economic violence perpetrated against people of color. It isn't just the current gang violence that destroys lives and shreds hope. It also has to do with land abuse embodied in the corridor itself, the trains that used the area up and then left it for better ports of call. The rail bed may now have a future as a literal lifeline, a taproot in the community that creates a different trajectory. Will residents embrace it? Will it really bring economic opportunity and jobs back to the community? Or could it, as some fear, be so successful that it becomes a tool for displacement, further marginalizing already vulnerable people?

I descend from the trail, on the way back to Elvia's parked car, and spot a Safe Passage sign near the street. These yellow street signs are spread throughout Chicago, intended to mark safe walking corridors for schoolchildren. Parents, staff from local community organizations, and volunteers take shifts along routes to make sure that children can get to and from school without trouble from gangs. The signs are Janus-faced, too. They signal community care, adults looking out for the next generation. But that signs are needed to mark the way to safe passage—that safely walking the streets in some Chicago neighborhoods cannot be taken for granted—offers a stark commentary.

Cities should be places of safe passage, for all people. For all creatures, really. When Safe Passage signs are no longer needed, a new era—not just a New ERA trail through the city—will have begun.

Ramble On

"On your left," someone hollers, and a few seconds later *whoosh*. At one point along the trail, deep in my own thoughts, a sudden "on your left" startles me into a karate-like defense pose. *Ki-yawp*. I'm certain I look like a complete doofus. Farther along the pavement, yet another "on your left" prompts me to fire back "on your right."

I'm on the Cal-Sag Trail, a newly constructed route that skims the southern outskirts of Chicago for twenty-six miles. Perhaps it's the return of life that a spring day betokens, but the new growth and the new trail feel like signs of a city waking. Change is in the air. The trail runs roughly parallel to the Cal-Sag Channel. One of many human alterations to Chicago's watershed, the Cal-Sag eased the shipping of industrial products and provided convenient waterborne disposal for industrial effluent. The steel smelting and heavy industry that made the area's name are mostly gone. Liberated from the burdens of heavy industry, the trail blooms and buzzes with life. The channel waters are, for the first time in a century, safe to swim (though one wouldn't want to nose around too deeply in the sediment). If the New ERA Trail hangs on the edge of possibility in a postindustrial Chicago, the Cal-Sag Trail represents a recovery already in the making.

The day offers the warmest day of spring yet, so cyclists have reason to dust off their spandex for a ride. There are fifty for every one person on foot I pass. I don't begrudge cyclists their time in the sun. All of the wheels whizzing by me, though, prompt considerations of the differences between walking and cycling—how the pace of the latter is more suited to where you are going than attuned to where you are.

In a city that still prioritizes getting from point A to point B in the quickest and most efficient manner, walking may be dismissed as an oddity or a hindrance. I recall one of the first big snows I experienced in Chicago. Walking to the elevated train in the early morning, I struggled over massive mounds of oil-stained winter slush on the sidewalks. The streets were plowed to pavement, the snow shoved atop the already deep accumulations on the footpaths, in clear demonstration of what modes of transportation the city favors to keep business humming.

Historian and writer Rebecca Solnit observes that the "apotheosis of speed," especially in the city, makes walking "a subversive detour." A detour, I think, worth taking; an act that will impact the future of what we demand of cities; an outreach that can connect us to a multispecies community. When a landscape is treated as empty space between a starting and ending point, we simply cannot know what a place has to offer and what we might exchange with one another. There is beauty to be discovered in embracing the inefficiency of the not-so-straight line. Walking "allows you to find what you don't know you are looking for," Solnit remarks, "and you don't know a place until it surprises you."[10] You put your body out there and see what turns up.

When our feet touch down, possibilities arise for new and needed ways of knowing that further knit the landscape together. Backcountry animist and beat poet Gary Snyder, in an essay entitled "Good, Wild, Sacred," shares a story about a time he spent with Aboriginal peoples during an extended visit to the central Australian desert in the early 1980s. He writes about the sacred songs and stories of that vast landscape—of how mythology, prominent natural features, and practical travel information are braided together.

In one scene, he and a Pintubi elder named Jimmy Tjungurrayi are traveling across a dirt track by truck. Jimmy points out important landmarks as they whiz by, describing their significance to his people, the animals they are associated with, and the elaborate cultural narratives embedded in them. Snyder notices that his companion's speech progressively becomes more rapid, so fast that he can no longer keep pace with the rich and complicated stories he is telling. This is when it dawns upon Snyder that these stories, among their many functions, also serve as mental maps for a journey on foot. "I realized after about half an hour of this that these were tales to be told while *walking*," Snyder writes, "and that I was experiencing a speeded-up version of what might be leisurely told over several days of foot travel." Barreling across the desert by truck, the vehicle's speed exceeded the stories' ability to be told or clearly heard.[11]

On foot, opportunities for discovery, for old stories and new, are easier to apprehend. Footfalls create a comfortable pulse that engages the body with its surroundings without overwhelming the mind with input. Thoughts can breathe. Walking is a procession that allows one to process.

In his book *The Old Ways*, Robert Macfarlane discusses several literary luminaries who credited their good thinking to their good walking. He names Jean-Jacques Rousseau ("My mind only works with my legs"), Friedrich Nietzsche ("Only those thoughts which come from *walking* have any value"), and Wallace Stevens ("Perhaps / The truth depends on a walk around the lake"). He also reminds us that the association between walking and thinking isn't simply the domain of Western intellectuals. Whole cultures—Cibecue Apache, the Thcho people of northern Canada, Australian Aborigines—describe their knowledge about how to live properly in terms of pathways. People walk to remember these connections, their bodies inscribed in a storied landscape. The link between understanding and walking has an etymological pedigree in the English language, as well; as Macfarlane notes, the Proto-Germanic root of the verb *to learn* means "to follow a track."[12]

By altering pace, a space is opened for encounter. Like the osprey pair I spot while on the Cal-Sag Trail. Osprey are impressive raptors known as "fish hawks" for their pescatarian diets. Their skill as experienced anglers can be seen in the way they carry prey gripped between their talons as they fly, with the unfortunate fish's head pointed forward to eliminate wind resistance. A rare state-threatened species, osprey have only a handful of nesting sites in Illinois. So, when I look toward a thin strip of trees overhanging the lazy waters of the Cal-Sag Channel and see not one but two osprey, facing each other on a branch, an involuntary yet entirely justified expletive escapes my lips. I absorb the sight—the white head with the distinctive dark stripe passing across a sunburst eye, the vulture-like hunch of the more than two-foot-tall body, the hooked beak that terminates in a needle designed to pierce flesh. Near the osprey pair, a jumble of sticks forms a large circular nest high on a wooden pole. A home for the next generation of ospreys. On the edge of Chicago.

Whoosh. A woman zips by me on her bicycle, braking slightly. "Is that a bald eagle?" she calls over her shoulder, the Doppler effect already dissolving her question. "Osprey!" I shout in reply. I don't think she hears me. She is already too far away.

A trail like the Cal-Sag forms part of a larger corridor, a threaded dyad of blue and green that passes like a stitch through the city's fabric. The defined edges of the Cal-Sag—the waterway on one side; a roadway, with trees muffling the sounds of distant automobiles on the other—forms a curving passageway through the built environment. A corridor of and for life.

Conservation biologists sometimes speak of the factors that contribute to a landscape's ecological viability as "the 3 Cs": (large) carnivores, cores, and corridors. Large carnivores because of top-down prey regulation, which buffers vegetation from the voracious appetites of herbivores. Cores because large contiguous blocks of habitat tend to support more species, some of whom need large amounts of unfragmented land to survive. Corridors because these are the arteries that keep the land's lifeblood flowing. Corridors provide safe passage—around, under, or over paved roads. Corridors offer escape routes and lifelines between populations of species that would otherwise be genetically isolated.

I want to continue walking this city corridor, but the time comes to reverse direction. On the way back, the trail stays superficially the same but new experiences alter its composition. One in particular stands out. Or flies out, rather. The band of trees on my left ejects a feathered blur across my path at eye level. Dangling from the talons of this now-ascending blur: three feet of snake. I arch my neck and fumble for my camera, hoping I can get a clear photograph, some proof that this is really happening. The red-shouldered hawk traces spirals in the air, crescents of translucent white shining through her wings, rising farther and farther into the robin's-egg blue of the sky. I can't tell if the snake is dead and swaying in the wind or very much alive and writhing against the forced trip away from familiar ground. The hawk recedes, circling toward a hill on the other side of the channel, and plummets out of sight. Only then do I realize my mouth has been open the entire time.

These trails, these corridors, are significant to other animals. As I put one foot in front of the other, it also occurs to me that trails such as these have significantly shaped how I've come to know the city as a living place, a discovery of the otherworlds of the urban wild. I sometimes imagine each footfall as a needle, me leaving a trail of stitches behind, threads of memory from experiences such as the hawk with the snake clasped in its talons. Every walk weaves me further into the land, joins my body's memories to the ground. Solnit writes that "walking is a mode of making the world as well as being in it . . . of knowing the world through the body and the body through the world."[13] There is a call and a response between body and landscape, which is just another way of saying between a body and the multiple other bodies that comprise a landscape. A body in place. A place in a body.

The pace of a city can constrain our vision of what a city is, what nature is, where each is allowed to be. Had I not been traveling by foot, I never would have seen the red-shouldered hawk with the snake in its talons, or the osprey pair, the tree swallows, the dragonflies, the great blue herons, the coyote scat, the rabbits, the red admiral butterflies.

Near the end of Snyder's essay, he writes: "The best purpose of [wilderness] studies and hikes is to be able to come back to the lowlands and see all the land about us, agricultural, suburban, urban, as part of the same territory—never totally ruined, never completely unnatural. It can be restored, and humans could live in considerable numbers on much of it. Great Brown Bear is walking with us, Salmon swimming upstream with us, as we stroll a city street."[14]

Or, in Chicago's case, Coyote is already walking with us, Catfish is already swimming up the channel with us, as we stroll a city street. Cities need not be defined solely by their tangle of pavement and buildings. That's only one story.

As I walk, I link paths by linking lives, human and nonhuman. The memories will be waiting when I come back. My feet will add further stitches to the fabric of the city. Weaving me into this place, weaving this place into me. The Chicago I've come to know on foot is an entanglement of life. An osprey pair living on the Cal-Sag Channel might deliver the same message.

Protecting walking paths, creating new ones, upholding the right of passage that corridors provide—all mark a contribution to the collective reweaving of the city. The greatest incentive for caring about such corridors is likely getting to know them with one's feet. The city has plenty of places to ramble and plenty of places that need ramblers, places to start and continue interspecies conversations, places to experience firsthand that nature is in and of the city. Given a slower pace, as well as renewed attention, memories, and associations, stories can emerge alongside new corridors. Maybe they've always arisen together in this way.

Blueways

> Gravity—I saw it in a flash—was the original sin, committed by
> the first living beings who left the sea. Redemption would come
> only when we returned to the ocean as already the sea mammals
> have done.
> —JACQUES COUSTEAU

Sinking

Mud slurps at my foot, quickly swallowing my calf, then my knee,
leaving me thigh-deep in viscous muck and ten yards from stable
ground. I cling to a double-bladed paddle and a hope that a 911 call
will not be necessary. Panic washes through my chest. In my mind's
eye, I watch news footage: a human chain of people, clasping hands
and reaching across a boggy mess to the frightened dunce whose
neck and head provide the only visible evidence of his woeful mis-
calculation. Onlookers fret and talk into cell phones about the sinking man. Several police vehicles and an ambulance, red lights flashing, lie in wait. The segment abruptly concludes with a moralizing

summation, delivered by an anchorperson wearing an expression of feigned concern: "Thanks, Susan. Whew. That was a close one. Rescuers say the man was badly shaken but unharmed. A good reminder that carelessness is dangerous in the great outdoors, even natural areas like the ones in our city." Twitter blows up with a gif of me slowly sinking. #DarwinAwards.

My mind refocuses on the present. Gelatinous goo oozes between my toes and I descend a few inches farther. *Is there a bottom to this?* I wonder. Who's to say how deep the goop goes—besides the white egrets, who ignore my plight, stepping gingerly atop the sludge with enviable stilt legs and hollow-boned bodies. My bones will be hard to find in a few minutes. I wish for a pith helmet—my cartoon-based version of what remains of a British explorer caught in Amazonian quicksand—so next of kin can locate me once I'm fully submerged. At least I will have saved them a trip to another continent. I am standing with one leg mired in a tiny tributary of the Chicago River, and the river is doing its best to absorb me.

Portaging

I began kayaking this urban river and its branches in earnest a few months ago with an ambition that this place would become more a part of me. I didn't literally intend to become a part of it. My urban kayaking adventures constitute an effort to see Chicago from a fresh angle, via its freshwater passageways. The water is the reason the city is here, after all.

Eighteenth-century trappers and traders needed an outpost that offered access to the interior of the continent, a waterway linkage between the Great Lakes and the Mississippi River. No continuously flowing corridor existed to bridge the subcontinental divide (given the modest elevation, the divide might more accurately be considered a subcontinental bump). A viable portage existed, however, between the Chicago River on the southwestern shore of Lake Michigan and the Des Plaines River, which flows into the hallowed intercontinental drainage of the Mississippi River and down to the

Gulf. Depending on one's seasonal timing, if the sluggish waters of the Chicago River were running high, a resourceful beaver trapper's feet might stay relatively dry.

Not everyone wanted to rely on the vicissitudes of seasonal conditions. Even Louis Jolliet (of the famous Marquette and Jolliet exploration duo), who crossed this divide only the once in 1673—and who opted for a dry portage path instead of a slog through a mile and a half of swamp—noted that a canoe-friendly canal would resolve the inconvenient separation between watersheds.[1]

I appreciate the temptation to alter the river's depth when I step out of my kayak into ankle-deep water on the Chicago River's North Branch, which wends its way through several forest preserves. Less a river than a creek here, the shallow idling water is unhurried, like an old man on a stroll. Fallen trees and limbs tend to stay in place for long periods of time. In a few spots, the accumulation of debris proves impassable, too much for me to get over or through. Around is the only way.

The debris is a chicken-or-egg situation. More humans utilizing this wisp of water would mean clearer passage, and perhaps clearer passage would mean more humans. As it is, I have the place to myself. And there are very few places in modern-day Chicago you can say that about. Being the lone person willing to maneuver around logjams peppered with plastic soda bottles and softened soccer balls, I am treated to a sense of *aqua incognita*.

Surprises greet me around every bend. A black-crowned night heron inquisitively eyes me from the cement tunnel of a sewage outflow. A doe and her fawn crane their necks upward from a shallow stream bank toward leafy tree branches above. A mama wood duck frantically flutters in an attempt to distract me from her chicks, who huddle under a decomposing wooden deck. Mud-stippled swallow nests burst with fevered *cheeps* beneath a highway overpass. A pair of golfers point and smile at the weirdo who is navigating their water hazard in a kayak.

And the water can be hazardous, I discovered, for both golfers and me. When I first pass between the manicured banks of the golf course, I increase my pace and try to reason with myself about the odds of a golf ball hitting my head. Such a scenario would require (a) a golfer with marginal-to-poor skills willing to fork over $100 in order to tear divots in a course for a couple of hours, (b) a shanked shot that goes terribly wrong but fails to strike the trees lining either side of the fairway, plus (c) the perfect space-time intersection between said shot and my unprotected position on the river.

Ploop! Against statistical probability, a small white orb smacks the water three feet from the port side of my kayak. I let out an involuntary *hoo!* of shock, then a *whoa! whoa! whoa!* for the benefit of any marginal-to-poor golfer within the vicinity, then paddle like hell to reach a bridge that will shield me from further aerial bombardment.

Despite the close call, this day's paddle is mostly uneventful, given over to the kind of reflections that surface so effortlessly on a river. Like the thought that there is a visible ethic of place embodied in the flow of a river. How people treat their water—as a precious force, an open sewer, an engineering problem, a hazard, a means of transport, a cause for celebration—speaks, without words, about the collective environmental ethic of a region. But that's putting it too academically. I'll try again.

How people treat their water tells you whether or not people know how to live for the long haul, whether they will adapt to place or not. Ignore the warnings—the stench of chemicals and feces, the sight of plastic cups and broken toys—at your own peril. If water is compromised, everything is.

For most of Chicago's history, people ignored the warnings. Heck, they actively exacerbated them. Bubbly Creek offers a good example of abuse. "Bubbly" because of the methane bubbles that rise from the decomposition of animal wastes below, the creek was a particularly heinous assault of sights and smells from the mid-nineteenth to mid-twentieth century. This little slip of water couldn't keep pace with the waste output of the Chicago Union Stockyards into the South Branch of the Chicago River. In 1911 a photograph was taken of a chicken performing an act of Jesus, walking on water so viscous it had become a solid without being ice.[2]

The Union Stockyards were only a single source of pollution, albeit a large one. At the time, no one seemed overly concerned with what was going into the river, whether human waste, animal offal, or chemicals from tanneries, distilleries, and glue factories. The solution for pollution was plain: send it downriver.

Reversals of the river occurred as early as 1848 and became a permanent feature of the watershed in 1900 with the completion of the Sanitary and Ship Canal. Ever since, the North Branch joins the main stem and both flow via the South Branch to the Sanitary and Ship Canal and, eventually, to New Orleans. For more than one hundred years, the waters of the Chicago River have flown away from the lake they fed for thirteen thousand years, since the recession of Pleistocene glaciers.

My office is in the Civic Opera Building, which sits on the banks of the South Branch of the Chicago River. Or it would if a riverbank remained. As in many portions of downtown Chicago, the river is lined with concrete here, the buildings plunging at right angles into water. On occasion, I look out my window and lose myself in the slow moving flow of the South Branch, knowing that it's going the wrong way.

I suspect many Chicagoans, even lifelong residents, are unaware of which way the Chicago River flows. Those who are aware probably share the view of the American Society of Civil Engineers, who, in 1955, called the reversal of the Chicago River one of the seven wonders of the modern engineering world. For my part, I have no interest in finding out what the other six are.

As my kayak drifts down the North Branch, a series of questions drifts into my mind: Can people of the city learn to align themselves with nature's cycles? Can we live as members of a larger-than-human community? Can we adapt to the way of water?

Following many local brainstorming meetings and decades of advocacy by Friends of the Chicago River, an aspirational document to guide recovery plans for the river was recently released. The Great Rivers Chicago plan includes goals to make the Chicago, Des Plaines, and Calumet Rivers "inviting, productive, and living." Among the goals are rivers clean enough for "primary contact" by 2020; a continuous riverfront trail by 2030; and by 2040, through "adopt-a-mile" programs, "native [species] communities will be abundant and healthy—some species healthy enough to eat for the first time since the 1800s." In the promotional lines of the document, one in particular stood out to me: "Ultimately, our rivers will define us, rivaling the lakefront in our hearts and minds, and become a key source of pride for metropolitan Chicago." It's a stroke in the right direction.

I haul my kayak ashore this day at a bend in the river that grazes against some open parkland in the forest preserves. I'm mud smeared, wet, and tired. It's been a good day. I flop down on the ground, using the inflatable kayak as a makeshift pillow. I watch a red-tailed hawk spiral above Chicago.

My thoughts follow the hawk's orbit, circling around the potential re-reversal of the river and what would allow the waters to join Lake Michigan once again in their ancient embrace. A couple of factors make this conceivable. Sewage treatment technology has advanced leaps and bounds; tunnels will be completed in the coming decades to better account for storm events that periodically dump polluted runoff into the river; and people are starting to recreate in substantial numbers on the river, increasing the demand for a healthy waterway.

For me, the re-reversal of the river would point toward a reversal of a water ethic (or its lack): from commodity to community. A commoditized object is on the business end of a monologue. We need not ask a commodity what *it* wants. A community participant, on the other hand, joins a conversation with others whose voices count. We are required to consider what members of our community want. It's worth trying to discern what water, the sustainer of cycles of life—and because of that, the sustainer of all conversations—wants.

Floating

Much has improved in the past four decades on the Chicago River. Better sanitation jump-started by federal water-quality standards, stream-bank habitat restoration, erosion control, species reintroductions. The river is clearly on the mend, gathering vitality.

But we're not yet within the cycle of water. When I paddle, signs peek from behind bankside vegetation to scream warnings in big red letters.

Caution

This Waterway Is Not Suitable for

- Wading
- Swimming
- Jet Skiing
- Water Skiing/Tubing
- Any Human Body Contact

The most troublesome bullet point is the last one. Kind of covers the gamut.

I'm on the banks of the North Shore Channel. Constructed in the early twentieth century, the North Shore Channel creates an aquatic link between the suburbs north of Chicago and the Chicago River. Basically the channel enlarged the piping of an expanding population's flush toilet. Yet a curious thing happened over time. With the gradual improvements previously noted, a considerable amount of the channel and the river have become unofficial wildlife refuges.

A gorgeous sunrise rouges the clouds over my shoulder. I unsling my inflatable kayak from my back onto the compacted gravel trail and set to work. The pining buzz of cicadas who have sung for paramours all night is punctuated by the intermittent trills of early rising grasshoppers. The morning air cools my forearms as I work the hand pump, inflating different pockets of what will soon ensure, fingers crossed, that "any human body contact" will be minimal. I'm after the spirit of the law here, not the letter.

I pull a few straps, turn the backpack inside out, and what was once the fabric shell I carried on my shoulders becomes the captain's chair that will carry me on the water for the next six hours. I click a few straps into place, Velcro the chair down, and pick up my aquatic transportation as I would a briefcase. Time to go to work. Not the most practical commute, but I got it in my head a few months ago that the sixteen-mile-or-so journey from northern Evanston, where I live, to the Loop in downtown Chicago would be a good way to experience the geographical and biological variety of the Chicago River.

Access points along the channel are scarce, but I know a place with a short climb over a chain-link fence. This leads to a couple of wooden stair steps that disappear into the water. I lift the kayak up and over, following after. The distant sound of cars sweeps through the tree canopy, and I slip into the kayak and push off. Local traffic report: only the honking of a few disgruntled Canada geese, who fuss at me and one another as they quicken their pace to widen the distance between us.

I notice the water flicking from the kayak blade onto my thighs and across my wrists. *Glad those caution signs are dated*, I think. I assure myself with off-brand logic that the sign didn't say anything about kayaking. All clear.

Great blue herons scold me with full-throated *waawk*s, voices reverberating like a wooden stick scraped across the ridges of a petrified gourd. My heart echoes against my rib cage in response. No matter how many times I accidentally flush a heron, I still startle every time they startle. Their pointed protestations tell me this is the one part of the city that they own, completely.

The herons seem flummoxed as to why this particular human decided to pass through their hunting grounds instead of sticking to the hard surfaces that other non-winged beings enjoy so much. Some are jumpier than others. The more seasoned great blues may be those that simply gaze upon me with indifference as I drift past. These have learned that the river is one of the few places they need not fear harassment, a tree-shielded corridor where humans walk a mere thirty, twenty, ten feet away, utterly unaware that a being with a two-foot-long neck and a six-foot wingspan prowls nearby.

Not only herons—kingfishers, flycatchers, dragonflies, beavers, mink, muskrat, turtles, frogs, and so many others share this thin ribbon of life. All are reminders that a city is only partially controlled, never completely regulated or under any one species' full jurisdiction—no matter what a politician promises. The river wraps itself through our best attempts at order, short-circuiting control with a living current.

Floating with the paddle at rest across my lap and my chin tilted toward the sky, I realize there's a reason I am in this kayak. I am here to breathe in original time. Rhythmic—heartbeat, tides, the lap of liquid wake upon solid stone. I come to the water to feel time as a circle that wraps around itself, not a line that extends into the distance. With a foot dangling limp off the edge of a kayak—caution signs be damned—all ripple lines become traces on a sphere. Move in a direction, drift with any flow, you become part of the orbit of water. Away from land, away from sure footing and so-called solid ground, time moves in spirals. On ground, it's easier to pretend that objects remain at rest; in the water, the jig is up. You are part of a flow—not motion across a static backdrop but part of motion itself.

Over the course of the day, I paddle the urban gradient, from hushed forest stream banks to attractive neighborhood homes to the corrugated steelworks of heavy industry. I drift by sewage treatment plants, cement check dams, the timber and iron ribs of railroad trestles, and under one-lane cart paths and eight-lane highways. At one point the Sears Tower emerges on my southern horizon, bolstering the energy in my flagging shoulders, although I seem to come no closer for all my efforts. At last I reach the confluence of the three waterways that constitute the Chicago River: the North Branch (which brought me here), the main stem (the short outflow that once meandered to Lake Michigan), and the South Branch (which now carries all these waters toward their modern destination in the Gulf of Mexico).

In the Loop of downtown Chicago, skyscrapers hug the banks on all sides. I inhale deeply and the kayak slowly spins on an invisible axis of competing currents. I feel as though I'm spinning forward and backward at the same time. Majestic new construction ascends, blue-tinted windows grabbing clouds from the sky, alongside landmark buildings like Bertrand Goldberg's Marina City "corncobs." History is jam-packed into this area of confluence.

The river's memory is old, witness to all. The buildings look eternal; the buildings look like a blip. I navigate around architectural tour ships, ride the wake of a couple of party boats, and salute a fellow kayaker or two. People on the banks gaze out upon the waters, eating their lunches, listening to music, absorbing the reflection of the sun. We've all come to be near the water. I think again about the city's relationship to the river. "Chicago created the river just as surely as the river itself was the genesis of Chicago," writes Libby Hill in *The Chicago River*.[3] We're in this together now. No turning back. The river rolls on.

There's something I'm trying to cup my hands around: how each generation must test the water worthiness of our cultural vessel, see and patch the leaks, put our faith in the river, and hope for decent weather.

Circulating

So many cities depend on dramatic manipulations of water. Chicago is not unique. Los Angeles hosts one of the more famous water diversion projects. Water moved to create a city forever destined to worry about water. The history of LA, writes David Ulin, is "a history of rapacious capitalism and vast infrastructure projects, of the men—and they were *all* men—who by their influence over various institutions (the *Times*, the Bureau of Water Works and Supply, the Pacific Electric Railway) used public resources for private good, building the city in the image of their greed."[4]

In the same time period that William Mulholland channeled the waters of the Owens River Valley into the Los Angeles Aqueduct to quench the thirst of a growing city's ambitions, Chicago faced a different sort of water problem: the need to move water *out* of the city. Why? Chicago didn't have a water shortage problem. Lake Michigan is the sixth-largest body of freshwater in the world. Chicago had a pollution abundance problem. The solution was not to move water in but to move it out.

Make no mistake, given the deplorable conditions of the Chicago

River, it was a service to public health to move the water on its way. (At least for Chicagoans; the St. Louis residents downstream didn't share this view.) Yet any time a people must physically move an offending substance away from where they live, they are likely focusing on the symptoms of a social problem rather than directly dealing with the disease. To move forward, we must learn how to treat the larger circulatory commons with more dignity. With love even.

Given an open heart, water is easy to love. I suspect most of us don't know why we are drawn to water. By real estate prices alone, an admittedly crude metric, we know we are. Nighttime satellite photographs show coastlines glaring like the rain gutters of neighbors who refuse to take down their Christmas lights. Vacations are planned for beaches and, if we can muster the funds, islands. Claude Monet spent a lifetime contemplating the play of light on water, the shifting moods of color; we spend time contemplating his contemplations, drawn even to representations of water. We attribute healing powers, legendary status, and sacred presence to water, from the grotto of Lourdes to the Fountain of Youth. These places make visible what we intuit: water is worthy of the divine.

Running water, given enough time, breaks down solids. The soft overcomes the hard, absorbing, transforming. Endlessly varied, endlessly cleansing—yet we still can't help ourselves from tampering, manipulating, controlling, channeling. Channelizing is embedded in the name of the North Shore Channel. Yet as I drift on this channel, below a sky full of swallows, I know that a name is not the whole story. A space has been created for water, and water, even water intended to flow as we see fit, retains a measure of wildness. Water is lifeblood, even to a city, especially to a city. It may be used to dilute unsavory things, but it naturally concentrates and distills life.

There's a book about flow dynamics that recently drifted into my life, though it was written fifty years ago. The title, *Sensitive Chaos*, grabbed me immediately. The book explains and illustrates what the author, Theodor Schwenk, refers to as "the archetypal forms in all flowing media," delving into the patterns and structures of hydrodynamics.

Before reading about the movement of water and how physiological structures have their genesis in flow dynamics, I assumed water cut its way from point A to point B, eager to incise itself in a landscape. But Schwenk corrected my misperception. He points out that water always tends to seek a spherical state: "It envelopes the whole sphere of the earth, enclosing every object in a thin film. Falling as a drop, water oscillates about the form of a sphere; or as dew fallen on a clear and starry night it transforms an inconspicuous field into a starry heaven of sparkling drops." He moves from these observations directly into the great circulations of water that break down and create life. "Water will always attempt to form an organic whole by joining what is divided and uniting it in circulation," he comments, nodding toward the phases of water as solid, liquid, gas; the circulating atmospheric currents that travel around the earth; the rising and falling of ocean currents. "Whether hurrying toward the sea in rivers, whether borne by air currents or falling to the earth as rain or snow—water is always on the way somewhere at some point in one of its great or small circulatory systems."[5]

Such circulations also happen at much smaller scales. A river, for instance, meanders as it does, swerving back and forth, oxbow after oxbow, because the spiraling flow of water. The downward pull of gravity straightens out what desires an endless curve. Every oxbow—resembling the Greek letter *omega*—is a relic, an arc of liquid entrapped and left behind, testifying by its shimmering blue presence to the trace of a circle's memory.

The natural course of a river is "rhythmic meandering," and when a river is confined or straightened, a vital energy is dampened. To use Schwenk's words, "Where it is deprived of rhythm and can no longer flow freely in meanders, or trickle over stones and murmur and chatter and form waves, it begins gradually to grow weary and die." Through this lens, the North Shore Channel was conceived to be stillborn. "A river that has been straightened out looks lifeless and dreary. It indicates the inner landscape in the souls of men, who no longer know how to move with the rhythms of living nature." True. Yet there is water there. Where there is water, there's a chance for life to assert a will of its own. "Not even the strongest walled banks can hold out indefinitely against this 'will' of the water," Schwenk remarks, "and wherever they offer a chance they will be torn down. The river tries to turn the unnatural, straight course into its own natural one."[6] Give it time.

I glance over my shoulder at the vortices left by my paddle strokes. With flow, the circle is pulled into a spiral. From antelope horns to snail shells to lungfish intestines, the spiral shows the organic arrangements of flow dynamics. The spirals are embedded within us. The human cochlea, "the vortex as an organic form at rest," part of our inner ear, is one example. Schwenk traces the journey of sound in the following manner:

> A glance at the external ear sees the spiral form winding its way into the internal ear and becoming lost there. Like a vortex with its funnel, the ear conch and the ear passage lead to the first membrane, the ear drum, against which the ossicles of the middle ear lean. Like a minute "system of limbs" (R. Steiner), they pass the rhythms they receive to the membrane of the fenestra ovalis in the internal ear, where they are led still deeper into the dark regions of the cochlea. The passage into the cavity of the internal ear is like a journey through the elements, from air via the solid medium to the liquid, and every form on the way reveals its origin in the archetypal movement of fluids.[7]

Vibrations on air create a current entrapped within our spiraling ear, registering on the drum, which gives them over to liquid, which gives them over to the electric pulses that travel neuron rivers to the brain, which interprets these sounds and crafts from them an intelligible language. Put your ear to a river. Liquid calls to liquid. Who says we can't hear what a river has to say?

In the case of Chicago, we can follow the spiral of the river forward and backward through time. Backward to industrial abuse, to a public commons becoming a public cesspool. Forward to a reclamation of vitality, a turning toward this common source of life. If the vision of the Chicago River laid out in the Great Rivers Chicago plan comes to fruition, it won't be too long—on the spiral timeline of a river's memory—until the caution signs are taken down and we'll be able to jump headfirst into the waters that brought Chicago into being, once again commingling with the many creatures who know the way of water.

· · ·

You might be wondering about the earlier episode, how I escaped the clutches of one of the Chicago River's shallow tributaries, because I obviously lived to tell the tale. With the paddle and kayak as leverage, I managed to extract my thigh-deep leg from the mire. Then, doing my best imitation of a white egret, I gingerly sloshed to a not-quite-but-solid-enough island of mud. After reembarking onto the shallow stream, I forded my way to the embankment of a drainage ditch and pulled my body ashore. I emerged, shaken and swampy, with not an inconsiderable amount of gratitude.

When I collapsed on the grass, the sun shone a little brighter and the breeze felt like a cool touch of forgiveness. The moral lesson, if there was one, was all's well that ends well. Or maybe look before you leap. Or maybe something about carelessness and the great outdoors. Or maybe something about how a river, no matter how controlled or seemingly innocuous, is a living, wild being. Fight against it and you may regret the error. Align yourself with it, adapt to its flow, and new worlds may open.

Mindways

At what moment, exactly, did the city of Chicago cease to be part
of nature? Even to ask the question is to suggest its absurdity.
—WILLIAM CRONON, *Nature's Metropolis*[1]

Boasting four million square feet of commercial floor space, Merchandise Mart was once the largest building in the world, large enough
to have its own zip code. A handful of meters south of this imposing art deco behemoth, eight large bronze heads perch atop a row of
evenly spaced pillars. The pillars line the north bank of the Chicago
River. These sculptures were commissioned in the 1950s, but they
honor famous retailers, all white men, who were born in the nineteenth century. Weathered green and streaked by pigeon droppings,
the heads possess an uncanny quality, severed at the neck, with their
gazes fixed upon the building entrance. Eight pairs of bodiless eyes
judging modern-day transactions, ensuring business continues un-
interrupted.

Although the heads had no choice in the matter, by virtue of facing
the building they also face away from the river, symbolically fitting

for a river that served as an open sewer during the era in which these men lived. It was a time when the flow of business—not of the river— was mistaken as the lifeblood of the city, and when, during these men's lifetimes, the river was engineered to flow backward, away from the city, to carry the froth of human-produced waste out of sight.

I wonder what these industrial-era titans, memorialized in bronze as retail conquerors, felt when they first looked upon Chicago. With no mountains to bend around, nothing but horizon except for the constantly expanding city skyline, the burgeoning metropolis must have seemed a blank canvas for unimpeded ambition. "What history Chicago did possess its residents commonly ignored because they felt little connection to it," writes historian Carl Smith, who adds, "Few places were so speculatively oriented. It is no coincidence that Chicagoans created the modern commodities market, in the form of the Chicago Board of Trade, where the future itself is bought and sold."[2] Economic growth defined the city's purpose—land as commodity not land as community—and stories were told of what could be, not what was.

The Puritans infamously claimed Boston as a city on a hill and a light to the Eastern Seaboard. Chicago offered a flat place in the midsection of the country to drive an axle into the ground and turn the gears of commerce. As those cogs rotated during the past two centuries, the city witnessed profound changes. From encampment site at the mouth of a modest river to ramshackle frontier town to industrial behemoth—no city grew faster. By the time the city had ballooned beyond two million residents in 1910, the unruly and traffic-jammed downtown streets produced a feeling that it was "choking on its own success."[3] Later came the escape from the city—the highways and the suburbs, the race to get away from other races, to leave behind the cityscape in favor of a lawn of one's own. Alongside cities in various parts of the world, Chicago today is entering a postindustrial phase. With manufacturing jobs lost to automation and cheaper labor available overseas, the city is deindustrializing. Now, as it has been since its inception, the city is reinventing itself.

This provides a moment of pause. The bronze heads outside of Merchandise Mart cannot turn toward the river, or lift their chins to the sky, or get on hands and knees next to the ground. But the rest of us should consider doing so.

There is a new story to tell about cities—one that can shift the plot from acquisition and exploitation to inhabitation. One in which the city is more than a means to acquire more. One in which the city becomes life affirming instead of life denying, a generator of biological complexity and diversity rather than simplicity and impoverishment. Such an understanding of cities requires a shift in perception. A summary of this shift in perception might be called an *urban land ethic*.

A land ethic, at least as Aldo Leopold phrased it, is deceptively simple: "A thing is right when it tends to preserve the integrity, stability, and beauty of the biotic community. Wrong when it tends otherwise."[4] A land ethic is also, he is quick to point out, never written, ever evolving, and finds expression in various practices that foster the long-term well-being and resilience of the land community. Leopold didn't hand down a chiseled set of well-defined prescriptions and proscriptions. There is no land ethic rulebook or step-by-step instruction manual. He instead offered an orientation, a way to think about land as a *social* relationship: "The land ethic simply enlarges the boundaries of the community to include soils, waters, plants, and animals, or collectively: the land."[5] He puts forward some interchangeable terms to capture the sense that we're all in this together: the indivisibility of the earth, the land community, the biotic community, and the collective organism, among others.

Collective organism is a phrasing that appeals to the imagination. A city is more than human, part of a pulsing earthly organism. It always has been. Embracing this as a reality requires alignment of human lives with the lives of nonhuman others for the good of the whole. The nonhuman animals among us have stories to tell about how we might move in that direction. The least we can do is listen.

Orienting toward the land would rescue the word *citizen* from its reduction to legal definitions, revivifying the roots of the term's meaning: *inhabitant* of a city. Alive in place. Not simply occupying urban space, a pixel on a screen, a billiard ball on a table, an interchangeable consumer in a Target, but a participant with one's fellow inhabitants in a common journey.

A voice in my head accuses me of being a dreamer. So be it. If we lack a dream, we can expect to arrive where we seem to be heading along the course laid out by the current story. That story is based on a fractional period of our species' existence—the industrial acceleration—approximately the two-hundred-year life span of Chicago. This narrative pledges its allegiance to a culture of invasion and acquisition. The financial benefits go to the individual; the social and environmental costs are absorbed by the collective—that is, other human beings, plants, animals, soil, water, air. The story tells a particular tale of what it means to be human, affirming the ideology that *it's simply what we do: tear things up, grab what we can, and move onward and upward.*

There are other ways to be human, to live with and not upon others, to honor our collective journey and cast our disproportionate vote among other species on the side of life. At least one piece of what I am advocating here is that other animals can help us think and behave differently. The human ecologist Paul Shepard reminds us: "The human species emerged enacting, dreaming, and thinking animals and cannot be fully itself without them."[6] To live into the fullness of our humanity, we need to dream, think, and act with other animals in mind.

Cities—their skyscrapers and roads, parks and department stores, art galleries and restaurants—provide visible testimony to the power of the human mind. We are a species that can conceive of something not there and, with access to the tools and resources, say, "Let there be," and make it so. Our species has become fairly sophisticated in identifying our own needs for housing, food, amenities, and all those things that make a place livable. We also can recognize when those needs are unmet or unfairly constrained by others. But the dynamism of the human imagination, the empathic reach of our thought, finds fullest expression in understanding and meeting the needs of those who are not human.

An increasing number of people recognize that ecological communities don't stop where the buildings begin. What happens in the city—including our attunement to the life around us and a heightened perception of our own creatureliness—can bring together in the mind that which is already connected in the landscape. Other animals allow the continuum of life to come into focus. This focus can be sharpened through small acts—micro-rewildings—at the scale of the backyard, the street corner, and the neighborhood. Thinking small allows us to think big.

. . .

From philosophers to zoologists, the point is often made that we humans don't have direct access to other animals' minds. This claim is frequently tethered to a healthy concern about anthropomorphizing. Attributing human motives to other animals carries the danger of collapsing important differences between species, thereby erasing their uniqueness for the sake of gleaning a moral lesson about our own. In the early twentieth century, accusations of anthropomorphizing reached a fevered pitch, and a heated row broke out that included several writers who were accused of "nature faking." The accusations hinged on personifying animals, particularly in stories presented as thrilling true tales.[7] Prominent public figures felt compelled to weigh in on the matter. Even then-president Theodore Roosevelt was miffed enough to throw his safari hat into the kerfuffle, accusing popular writers of the day such as Ernest Thompson Seton of misleading "facts" and poor scientific acumen.

While that debate waned long ago, some of its aftermath still exists. Careful avoidance of anthropomorphizing, especially when scientific authority is at stake, often translates to treating other animals as "its," assigning numbers not names, and maintaining a stark emotional distance for the sake of objectivity. Appropriate perhaps, under limited circumstances, but this distancing has consequences. Infantilizing other animals—stuffed plush toys, Disneyfication, cartoon bears, and costumed pets—is generally accepted; taking seriously the lives and agency of wild animals, respecting their subjecthood and other-than-human intelligence, is less common. Projection, people say. Kids' stuff. Grow up.

We don't have direct access to animal minds. This doesn't bother me. We don't have direct access to anyone's mind. My son's behaviors are sometimes as unintelligible to me as a butterfly's. What we do have is imagination and the power of observation. This combination may be one of the human species' greatest graces, and it leads us toward other species not away from them.

Thinking small can lead to thinking big. Thinking like a bee, for example. A handful of people dared to do so, and now the street corners of one Chicago neighborhood are abuzz with life.

I'm standing on one such street corner with Lisa Hish, near her home in Chicago's North Center neighborhood. A car pauses at the four-way stop, then putters past. It's easy to discern something different here from the standard city flora. The turfgrass and pavement that populates so many curbs throughout the city is ceding ground. Instead, plants such as columbine, butterfly weed, and marsh blazing star whisper in the breeze. Other touches—a small birdbath, a stone pathway—indicate that these corners are the result of intentional acts of cultivation and care.

Lisa catches me up on what's going on. Multiple strands combined to form what she and her neighbors call the pollinator path. One strand began with the Parkway Corner Initiative, fourteen corners cared for by one to three neighborhood families who pledge to maintain them for three-year stints. The initiative began as a two-year pilot program, the brainchild of Laurel Ross of the North Branch Restoration Project and the late Elizabeth Wenscott, who was Lisa's partner, a member of the Northcenter Neighborhood Association, and a master instructor at the Tai Chi Center of Chicago. The rules are simple: more native plants, no pesticides. After that, do what you like to make your corner plot as attractive as you can for bees, birds, butterflies, and even bats. I put my nose to the fuchsia petals of a bee balm plant and inhale. I wonder if bees can feel joy, if various nectars are as seductively heady as fine wines.

Lisa interrupts my bee balm reverie. Another strand contributing to the pollinator path, she tells me, has to do with city living. "Some of us have the good fortune of living near mountains or redwoods or can afford to visit them. But what about those who cannot? And what about those of us who need to be reminded of our connection with nature in daily life?" she asks. Even though Chicago has some prime green spaces, oftentimes the homes near those spaces come with a hefty price tag. "This makes yards and our common areas more critical," Lisa observes. "In some building-dense areas, the parkways may be the only common green space. Can we afford *not* to cultivate them?" Hard to argue.

The final strand is more personal. After the death of their good friend and beekeeper Lora Krogman, Lisa and Elizabeth acquired a beehive for their yard. This swarm of four-winged neighbors led to reimagining the neighborhood as a landing strip for pollinating insects. In an effort to ensure that the bees would be happy with their foraging prospects, Lisa and Elizabeth created a map and began requesting that their neighbors mark this map with any bee-friendly plantings occurring in their own yards. The idea of actively supporting bees and other pollinators spread among neighbors, and so did the native wildflowers.

This led to still-larger plans: the notion of a full-blown pollinator path. The big picture involves sprinkling patches of attractive pollinator habitat through various neighborhoods, creating a linear corridor that reaches from the Chicago River in the west to Lake Michigan in the east. Although a solid line of flowers and native greenery is a present impossibility because of the pavement, Lisa and her co-conspirators in reclamation see opportunities awaiting in every yard and on every street corner.

Led by Lisa's breathy shih tzu–poodle mix, Mochi, we stroll through the neighborhood, stopping at various intersections that reveal the fruits of the Parkway Corner Initiative. When Mochi pauses to sniff a prairie dropseed plant, I ask about the way these lively patches of green have affected Lisa's view of the city. "I used to think of cities and urban areas as granting us proximity to live and work with easy access to each other and the commons," she replies. "But I was defining the commons as services: libraries, public transport, parks, blocks of businesses." A subtle shift occurs in my vision as

Lisa's words drift through my head. Like seeing a book's white space instead of the ink that forms words. The scrawl of the cars and roads fades; every lawn, every road median and shoulder, every balcony and flower bed and roof and forgotten patch of grass between buildings emerges from the background as an opportunity for life. "Now I see cities as a habitat," she continues, reaching down to scratch Mochi's caramel-and-white tufted head, as "a place to keep us living efficiently as a species, but also places where we still need to be in relationship with other species, for our health, our understanding of our place in the world."

Our place in the world. Perhaps that's easy to forget in the city. There are a slew of studies that show the important role that greenspace plays in human psychological and social well-being. People intuitively take their children to parks, hang out under the shade of trees, go for longer walks near the river or by the lake when they need to declutter their heads. Cultivating relationships with other species can puncture our focus on human affairs and open us to a larger community of life. Fortunately, if one takes a few steps away from the pavement, that life can be found almost anywhere. Or, in the case of the pollinator path, directly adjacent to the pavement.

During the course of our conversation, I discover that Lisa is a tai chi chuan practitioner and deeply interested in bringing body and mind into balance. As Lisa speaks, the connections between tai chi practice and the pollinator path assume greater clarity for me, as well as how the greening of this neighborhood represents an extension of her own life path. Lisa and her partner, Elizabeth, cofounded Elizabeth's school, the Tai Chi Center of Chicago, with Elizabeth as the master instructor and Lisa leading the health and wellness program. They later added to the school an environmental arm called Sustainable Return, where Elizabeth led the native plants and pollinator program and Lisa headed the food sustainability and vegetable gardening efforts. Many volunteers in these programs lent their helping hands to the pollinator path.

A lineage-based tradition, the style of tai chi taught by Elizabeth includes sixty-four postures that students must learn. They all rely on body awareness and positioning oneself in space so as not to over-extend or underexpress one's movements. The aim is stability and efficiency of motion, and while its applications are relevant to martial arts, Lisa continues to understand her practice as the source of larger lessons that can be applied in everyday life, particularly in the care of one's surrounding environment.

The teachings of tai chi chuan, Lisa tells me, are mirrored in a concept that Elizabeth taught: "knowing your size." My mind wanders to ecological footprints and overshoot, the kinds of quizzes I used to give students when I taught environmental studies classes that measured how many planets would be needed if everyone lived as most of us do in the United States. Those quizzes were always eye opening and more than a little disturbing. Knowing your size, Lisa says, means occupying your space but not the space of others unless that is intended. It's a teaching that extends to each person's relationship with the natural world. For Lisa, the pollinator path expresses a respectful occupation of space, welcoming other beings into the worlds that we've created in the city. "If we don't practice coexisting now with seven billion of us, how will we do so later?" Lisa ponders aloud. "Our neighborhoods are ecosystems that include us."

We circle the block, Lisa waving at people we pass along the way. The understanding that neighborhoods are ecosystems that include us seems to be spreading in our immediate vicinity, no doubt in part because of the colorful visibility of the Parkway Corner Initiative. Lisa tells me about other acts of micro-rewilding among her neighbors. There's the story about a summer evening when she was walking Mochi at twilight and met a man who casually told her he had spotted a coyote a short way down the block. When Lisa turned the corner, sure enough, she caught a glimpse of a coyote trotting quickly toward a side street. "Without a gait change, the coyote looked me in the eye while passing before turning into the alley." But when she recalls the encounter, it's the man's demeanor she most vividly remembers. His tone was not one of fear or "a call to mobilize" but a casual statement of awareness. He was "merely communicating information." No hysterics. Coyotes, at least for this neighbor, were an accepted, even normal, part of neighborhood goings-on. Lisa believes this represents a deeper change in her community about attitudes toward wild animals: a shift from fear to curiosity.

The wilding is spreading. The neighborhood now includes three yards that are nationally certified wildlife refuges. Rain barrels and bat houses are appearing in other yards, and pockets of turfgrass and ornamental bushes are being replaced by pollinator-friendly forage and prairie plants. Such transformations signal a nascent community identity. "Neighbors are more freely expressing and acting upon concern for the wild creatures within the area," Lisa notes. "For some, our peaceful coexistence has become a point of neighborhood pride."

Occasionally local wildlife literally brings people together, as when a small, fascinated crowd gathered to watch an owl and hawk contend with the harassment of crows. And there are the small daily moments, Lisa recalls, such as when parents and children find tufts of fur and scattered feathers on the hoods of their cars and underneath trees. The children can hold in the palms of their hands evidence of the cycles within which we all exist.

And of course, there are the corners, buzzing with life. The pollinator path is a vision in progress, as all works of the imagination are. The first of three segments is complete, extending along the appropriately named Grace Street, from the Chicago River to Lincoln Avenue. The next two segments will lengthen the pathway until it reaches Lake Michigan. Lisa's already thinking ahead: once the pollinator path is complete, she's hoping another two pathways will be built through nearby neighborhoods, providing further linkages between restored riverbanks and bird sanctuaries on Lake Michigan.

No, humans cannot literally think like a bee. Nor can we see in the ultraviolet spectrum, which no doubt colors the thinking of bees. Nor do we do a waggle dance to communicate where a food source may be—unless we're in a good mood. But we don't need access to a bee's mind to think about what a bee needs, and that kind of empathic thinking can be transformative. Thinking like a bee, in this sense, means thinking about what a plant desires. Thinking like a plant, among other things, means thinking about sunlight, water availability, and soil quality. Tug at a lifeline and the thread inevitably leads to an entanglement we share with every creature. Call it ecology if that suits you. Call it community if that sits better in your ear. Both words make the same claim: we are connected.

As we head back to her home, Lisa reflects on the many instances where neighborhood chatter about removing animal "threats" has become a discussion about alternative possibilities. "I think we stop the conversation and potential problem-solving too quickly when we use the mind-set of conquering-domination versus coexistence," she says. With a different set of words, she's just summarized Leopold's call for humans to behave as plain members and citizens of the biotic community. In his words, "A land ethic changes the role of *Homo sapiens* from conqueror of the land-community to plain member and citizen of it. It implies respect for his fellow-members, and also respect for the community as such."[8] I mention this to Lisa as she's opening her door, and a grateful Mochi scampers inside. Lisa turns, a twinkle in her eye. "Knowing your size," she says.

■ ■ ■

Ethics can be a fearful word. It sounds heavy as an iron doorstop. Authoritative as a judge's gavel. Ponderous as a classroom lecture. Oftentimes *ethics* is taken to mean a *personal* set of guidelines, an individual's code of conduct. More helpful would be to regard it as a collective effort. I imagine ethics as approximating another E-word: *empathy.* Other-centeredness. Lines of life reaching outward to embrace others' cares and concerns as worthy. In this rendering, ethics is a verb, a collective practice, that begins with listening to and with others, including nonhuman others. It's a relatively new thing under the sun for Western philosophy to consider nonhuman animals as ethically relevant. But a relatively old thing in the span of human history.

Perhaps because I walk physical corridors in the city, my imagination easily slips into thoughts about the corridors of the mind, the neural networks whose far reaches taper and branch like mycelium. These nerve endings, the dendritic root systems of the brain, strain toward one another, our thoughts and memories reconfiguring their pathways, forming bridges, and reinforcing the electric pathways that animate our constantly exploding minds.

Even when we sleep—especially when we sleep, perhaps—our brain is rearranging these meshworks, as though playing within an M. C. Escher drawing in which the staircases twist back to other staircases, shifting and bringing subconscious ideas up to consciousness and letting fizzle what no longer grasps our attention. I look at maps of the earth and think of the places I haven't been and want to go, but the brain contains territory as unknown as the Mariana Trench. And it is reconfiguring all the time, a room with doors leading to brick walls that you later open only to find spiral staircases to destinations unknown.

New patterns of thought about the purpose and possibilities of a city can create new corridors of life in the urban landscape. Imaginative leaps across neural bridges may build the bridges between our lives and those of other creatures, and may compel us to demand corridors that repair the frayed weave of life-giving pathways throughout the city. Rewilding the mind can rewild our cities.

When we ask ourselves, *How can wild nature be better integrated with our urban places?* we are also asking, *How can we, as humans, be better integrated into this place upon which we depend?* For answers, we would do well to look to other species for prompts about how to reenvision the city, to rethink its purpose, to reconsider its infrastructure, to reconcile our lives with the lives of those who share this place and those who can and would if given the chance. Sometimes this will mean acts of preservation, stepping back so others can have the space they need; sometimes this will mean restoration, stepping in to change the landscape and unearth time-honored connections; sometimes this will mean reconciliation—knowing our size—stepping forward to adapt and adjust our lives so that we can live alongside one another. We need every method to fully express an urban land ethic.

I would be remiss if I didn't note that an urban land ethic faces some inherent obstacles. In cities, our dependence on the land may be more shrouded. As a result, our imaginations must be larger and our education more intentional to trace the linkages and internalize the meanings between where we are at the moment and the places from which we get our food and heat and water.

In a city, our attention may also be hijacked by a near-constant stream of sensory input. We can both insulate and distract ourselves with greater ease. Yet compassion grows just as well in an urban human heart as it does a rural one. The question is not if we will direct our attention toward things other than ourselves but *to what and to whom* will we direct our attention?

I have suggested that other animals allow us to see ourselves and the prospects of the cities we live in more clearly. Their stories are entwined with our own. We need their continuous reminders that this is a shared journey. An urban land ethic, simply stated, is a practice of presence and care—intentionally seeking out what nonhumans tell us about where we live and responding by tending to the biotic community we share. An urban land ethic consists of the small acts that occur in the backyard, rewilding practices at the neighborhood scale, and, if we are privileged enough to be in a position to do so, the large acts that make a city a place of flourishing within our larger bioregion.

There are times when the forces levied against the flourishing of life seem too much to resist. In the United States, plenty of evidence exists to support a claim that we have not yet transcended a frontier-based psychology. The dominant narrative remains: the individual as an isolated conqueror against the world, rushing to carve a slice from the pie before someone else does, gathering a harvest he did not sow, and sharing only when compelled. The task of reversing that cultural mentality is daunting.

"I have no illusions about the speed or accuracy with which an ecological conscience can become functional," Leopold writes. Despite the prospects, he exuded a pragmatic doggedness about the task: "In such matters we should not worry too much about anything except the direction in which we travel. The direction is clear and the first step is to throw your weight around on matters of right and wrong land use."[9] Over 2,500 years ago, in one of the more famous lines of the *Tao Te Ching*, Lao Tzu said something similar: "A tree that fills a man's embrace grows from a seedling / A tower nine stories high starts with one brick / A journey of a thousand miles begins with a single step" (verse 64). No matter if delivered from a wizened Wisconsinite or a Chinese sage, this remains good advice for the road toward an ecological conscience. Take a single step, then another.

To take those steps, we need our fellow wildling kin. Other animals turn our heads from the mirrors of polished steel, jostle old memories, deliver the antidote for our self-imposed amnesia. Since the dawn of our own species' emergence we have been in their company, become who we are alongside and with them, known how to behave by looking to these kin for cues. A fuller humanity, a greater personhood, is developed by attending to these others. They break the cubicle, spiraling around a question like a peregrine falcon: What does this patch of earth require of you for *all of us* to live well? They are our relations and their eyes are upon us.

Nature in the city for me is downy feathers floating like summer snowflakes from the nest of a Cooper's hawk. It is the wild-garlic aroma of a striped skunk drifting through the neighborhood. It is the heron's *quork* of startled surprise before lifting skyward. It is the taste of lake water at the edge of my lips. And, yes, it is the flash of peppery, rufous fur that could belong only to a coyote dissolving into shrub.

A city can be a vast *taking*. An ecological conscience demands it be reimagined and rearranged as a vast *giving*, responsive to the lands, waters, and skies that supply its lifeblood. Blueways, greenways, mindways—corridors that connect what is in isolation, from life-giving habitat to freshets of the imagination. Take a single step, then another.

Postscript to a Hope

"You can't smoke in here!"

The man at the entrance pauses, door still ajar, late-afternoon light slicing into the room. Startled, he looks over his shoulder to see if the barkeep intended the command for someone else. Eyes widening in embarrassment, he realizes the proof of his guilt dangles from the corner of his mouth. Quickly he snatches the bowl of the pipe and tucks it into the front pocket of his button-down short-sleeved shirt.

The shirt is festively patterned, suggesting a tourist on vacation, fashionably at odds with the spectacles on the man's nose and his creased khaki slacks. Perhaps in his sixties, his hair silvered and thinned, he looks like an undercover professor. If professors ever needed to be undercover.

He scans the tables in the barroom, squinting into the dimmed interior, on the lookout for someone. A bar stool props up a patron by the end of a long wooden countertop. The man's eyes tighten on the patron, the one with hairy—furry even—ankles visible between the gap where his pant legs end and his shoes begin. A hint of a smile crinkles the man's face. He steps confidently toward the patron.

"What took you so long?" Coyote whispers conspiratorially, turning sideways on his stool and knocking back the remainder of his beer.

"An awfully slow bus ride," says Aldo. "A lot of cornfields." He takes a paper napkin from the countertop and gently dabs the sides of his forehead, then orders a bourbon.

He makes mental note of the row of empty beer glasses. "Sorry to have kept you waiting."

Coyote also makes mental note of the row of empty beer glasses. "No worries." His top lip curls, canines reflecting against the pint glass. "Let's go find him."

The companions exit into late-afternoon daylight. Coyote kicks off his shoes and peels his coat and pants off. "No need to keep up formalities now that we're outside," he explains. Aldo notices, though, that Coyote fails to remove the radio collar from his neck.

The buzz of spring is palpable in the city air. People stream down the sidewalks. Aldo turns toward Coyote to remark on the small white flower, draba, that has shouldered its way through a crack in the pavement. But Coyote is gone.

The flash of a black-tipped tail moving between a blur of legs catches Aldo's eye. He follows. Occasionally, he loses sight of the tail, but he is unworried; walking east, they will both eventually end up at the lake. With the river on his left-hand side, before reaching State Street bridge, Aldo spies his mischievous companion. Coyote moves as though he is following someone, too.

A businessman pauses at a crosswalk, waiting for the signal. He is talking to the air. No, not air—talking to a small device hanging from wires connected to his ears. Coyote approaches and slips the radio collar from around his scruffy neck and buckles it to the ankle of the businessman. The light changes and the man walks away, none the wiser.

Aldo catches up to Coyote, who waves a paw at the businessman. "Buh-bye," he chuckles, cocking his head at Aldo. "Let the scientists figure that one out."

"You haven't changed a bit," says Aldo. The companions turn and descend a flight of stone steps toward the river. A cement path hugs the water, leading under a series of bridges. They follow the path. Other people passing by, preoccupied, don't seem to notice the pair. Or perhaps their going unnoticed is a trick of the light. Above them, buildings dazzle in the sharpness of the spring sun. The air feels charged by springtime awakenings—from tiny violets to potted trees, everyone stretches toward the warmth.

A barge carrying construction equipment rumbles slowly along, passing a tour boat going the other direction. Encouraged by the amplified voice of their guide, people on the upper deck of the tour boat crane their necks in search of the "Bauhaus style." Aldo, too, ponders the skyline. Coyote follow his gaze to a few buildings whose roofs reveal the leafy crowns of trees. They both appear amused.

"What's the point of putting the trees up on the roofs?" wonders Coyote, ears twitching.

"There is value in any experience that reminds us of our dependency on the soil-plant-animal-human food chain," answers Aldo. "Civilization has so cluttered this elemental human-earth relation with gadgets and middlemen that awareness of it is growing dim. We fancy that industry supports us, forgetting what supports industry."[1]

"A peculiarly human problem," Coyote offers.

"Afraid so."

"Put a tree on every roof if it helps," Coyote says, still staring up at the green roof. He drops his eyes and scans the area, sneering at the pavement. "May want to leave some on the ground, too."

"Can't argue with that," says Aldo. "Twenty centuries of 'progress'"—Aldo curls his fingers, making air quotes—"have brought the average citizen a vote, a national anthem, a Ford, a bank account, and a high opinion of himself, but not the capacity to live in high density without befouling and denuding his environment, nor a conviction that such capacity, rather than such density, is the true test of whether he is civilized."[2]

"That's a mouthful, professor," scoffs Coyote. Aldo smiles and shrugs: "Hard to shake old habits."

They walk farther east, nearing the lake. Gulls circle and cry. They pass a group of plump pigeons who are bobbing their heads, tearing at a discarded sandwich bun. Coyote licks his lips. "Later," cautions Aldo. "Don't ruin your dinner."

. . .

As soon as the pair reaches the lakeshore, Coyote lifts his shiny black nose to the air and sniffs, closing his eyes in ecstasy. "Fish," he says, as though the word is a prayer. They spot an older man in robes who sits beside a small campfire, his long, gray hair loose and tumbling across his shoulders. He gently rotates a pan on top of a grill. Coyote trots over to the fire. A small fish crackles in a pool of butter inside the pan.[3]

"I thought you'd show up soon," the older man says, without lifting his head.

Coyote extends an arm for the fish. *Smack!* With a quickness that belies his age, the older man deflects Coyote's paw. Coyote gives a yip—surprised rather than pained—and glowers at the man, shaking his paw. "Well, then," he says, trotting to the other side of the pan, outside the reach of the man's arm. The man remains still, watching the fish sizzle. This time, Coyote reaches out, touches the fish . . . and then yowls in agony. Sprinting for the lake and clutching his singed paw, he jumps headfirst into the water with a mighty splash.

The two men on the shore attempt to stifle their laughter—but can't. They burst into a series of guffaws.

Relieved by the cool water, Coyote takes his time swimming to shore, scraggly haired and soaking wet but no worse for wear.

"I have a feeling you knew that would happen," he says when he paces close to the old man. "How about this?" He rapidly twists his body, sending sprays of water onto the fire, onto the man, onto the fish.

All at once: the laughter stops, the fire goes out, and the man releases the handle of the pan, stumbling backward. Coyote grabs the fish and swallows it in one gulp. With a deliberate glare from his piercing yellow eyes, he looks first to Aldo, then to Lao Tzu. His face softens and he flashes a big toothy grin. In unison, all three companions break into laughter.

When the laughter subsides, they sit down on the soft sand, relaxing into the evening's peace. A gentle breeze ripples from the lake. The distant honks of drivers commuting home mix with the honks of Canada geese and fade into the water, swallowed by the waves.

Coyote breaks the silence. "So you've come back," he says, turning to Lao Tzu. "I've always wondered, why'd you leave in the first place?"

"Oh," the old man sighs, as though he had never considered the question. "It was time." He stares up at the sky, contemplating. "The Sage shows people how to be simple and live without desires. To be content and not look for other ways. With the people so pure, who could trick them? What clever ideas could lead them astray?"[4]

"I could trick them," Coyote retorts.

"Not even you could trick people if they were truly content."

"Seems they weren't," Coyote says, gesturing toward the buildings and the roadway at their backs. "My hunch is you just wanted to see the mountains. The concerns of the city faded with each hoof clop from the horse you rode out of town."

Lao Tzu ponders this, inhaling deeply: "I do admit, when I swept the brush from the bamboo and laid it aside for the final time, the soft clink of the handle against the inkstone satisfied me. I thought it was the last time I'd hear it. I rose from my cushion with a single thought in my head: *done*. When I left, I was a soft breeze passing through gingko leaves. The gatekeeper didn't even notice my departure."

The memory seems to transport his body back to that time. Lao Tzu stretches his thin hands high, until his back pops, then rests them in his lap with a sigh. "How's that line go," he says, tapping his chin. "Ah, yes: The Sage acts without action and teaches without talking. He works but not for reward. He completes but not for results. He does nothing for himself in this passing world, so nothing he ever does passes. So I went. But the work lives on."[5]

Coyote looks puzzled. "So why are you here now?"

"For similar reasons as you, my friend. I was needed. People have not lived in accordance with the nature of things. In truth, there are many more people and the grabbing and stuffing—there is no end to it."[6]

"True enough," interjects Aldo, who had been listening quietly. "The shallow-minded modern who has lost his rootage in the land assumes that he has already discovered what is important; it is such who prate of empires, political or economic, that will last a thousand years. Most civilized men do not realize that science, in enabling us to see land as an organism, has given us something far more valuable than motors, radios, and television."[7]

"I don't need science for that," says Coyote. Lao Tzu chuckles.

The sun begins to set, although the buildings to the west make it impossible to see its last dip below the horizon. The shadows lengthen across the lake, pooling together. Aldo scratches the top of his ear and presses his hair back with a free hand. He tilts his chin toward the sky. The companions see the first of the stars winking over the lake.

"That's something, isn't it? You can see the Big Dipper. Just over there," he points.

"I've got a story about how it got there," says Coyote.

"You have a story for everything," teases Aldo.

Coyote shrugs. "I get around."

Aldo concedes with a quiet nod. Then, staring at the stars, he waxes philosophical: "Possibly, in our intuitive perceptions, we realize the indivisibility of the earth and respect it collectively not only as a useful servant but as a living being—a being that was old when the morning stars sang together, and when the last of us has been gathered unto his fathers, will still be young."[8]

"This is Tao!" exclaims Lao Tzu, his placid expression replaced with a thin smile.

"What the hell is Tao?" asks Coyote, adding, "Something edible, I hope."

"Something formless, complete in itself. There before heaven and earth. Tranquil, vast, standing alone, unchanging. It provides for all things yet cannot be exhausted. It is the mother of the universe. I do not know its name. So I call it Tao. Forced to name it further, I call it *that to which all things return*."[9]

"So *not* something to eat," Coyote responds, dejected.

The darkening sky over Lake Michigan becomes nearly indistinguishable from the horizon band of lake water. Lao Tzu picks up a handful of sand, letting it pour through his fingers: "The world is nothing but the glory of Tao expressed through different names and forms."[10]

"But is it real?" asks Coyote.

"It is the spirit, the essence, the life breath of all things. 'But is it real?' you ask—I say its evidence is all of creation."[11]

Coyote raises himself up and begins running in circles, kicking sand on his companions. "Well, then, let's go find it everywhere. Let's run. Let's swim. Let's throw ourselves against the sky. Or, at the very least, shoot arrows to make a ladder to the stars. I have old companions up there."

Lao Tzu closes his eyes, takes a deep breath, and opens them. His pupils are unfocused, resting on a point far out from shore. "No need, friend. Every place in this world is the perfect place to be."[12]

But this place would be really perfect, Coyote thinks. *If we had more fish.*

▪ ▪ ▪

The three leave the campfire and begin strolling the borderline between city and water. Many stories are exchanged along the way. Stories of the ten thousand things, stories of stealing fire, and stories of mountains that can think.

Coyote perks up. "After me, someday, there might be space in the city for Wolf, Mountain Lion, and Bear," he boasts. "At least to come nearby. Then be on their way. I've got things covered here."

"Wolf, huh? That'd be something," says Aldo. He reaches into the front pocket of his shirt, fetching his pipe and pressing it to his lips. He strikes a match, shielding it from the wind with the cup of his hand, watching as the ashes glow green in the bowl. "If that happens, they'll have you to thank for blazing the trail." Aldo flashes a sidelong eye-roll at Lao Tzu.

"I suppose most two-leggeds won't care much, one way or the other," says Coyote, ears twitching, "but I'll be poking around the edges, alerting the ones who are willing to hear. None of us is meant to sit on a throne. There are many around the table. Time for them to have their seats back."

With this declaration, Coyote raises his head and proclaims: "I'll be haunting this place until humans are willing to listen: the lines between our lives are muddy. You can re-create how things fit together—I know this! You can build, or not, with minds turned toward your animal-kin in the city—I know this! You can create new paths and not destroy the old ones—I know this!"

Lao Tzu ponders Coyote: "You'll have to keep reminding them."

"Yep," Coyote agrees.

Aldo puts a hand on Coyote's shoulder. "And remind them again."

"Yep," Coyote agrees.

A long silence passes among the three.

"Or not," Coyote says.

Aldo and Lao Tzu stare at Coyote in stunned silence.

"But . . ." Aldo stammers, not quite able to get the words out.

Lao Tzu sighs and shakes his head.

Coyote winks.

"Only kidding," he cackles. "It's what I do."

"Till another time, friends," he exclaims, and bounds down the shoreline, kicking sprays of lake water and sand behind him, swerving in large S's, following the edge of the tide while the waves hungrily lap at the shore and recede. The human pair watches his erratic gait as he prances away. Soon Coyote is merely a dark silhouette in the distance, then a barely discernible spot in the city twilight, ghosting to a vanishing point. The remaining companions turn from the lake, back toward the city. The lights from office windows blink on, others blink off. Red and yellow blips pulse from antennas set against the navy-blue night sky. A cool wind rises for a moment—sweeping sand in small eddies across the shore, rustling their clothes.

"Follow your path to the end," Lao Tzu whispers in the direction where Coyote disappeared. "Accept difficulty as an opportunity." Then, barely audible, so that the wind seems to dissolve the trace of the words: "This is the sure way to end up with no difficulties at all."[13]

. . .

If you had come to the lake that evening to walk the shoreline—to breathe deeply and cast off the troubles of the day—if you had been there in that moment, looking at that precise spot on the shore where the three companions gathered, you may have seen a shimmer of something, a patch of sand flickering with shadows, though the moonlight would have made you second-guess yourself.

But unmistakable in that evening air, with the traffic muffled by the lake's murmuring, you most definitely would have heard a defiant cackle-yip, a song arriving simultaneously from far away and some place very close.

GRATITUDES

This book wouldn't have happened without the sustained support of a beautiful community.

To my family—Marcy and Hawkins, who hold my heart. Three is the magic number.

To my Center for Humans and Nature colleagues—Jim Ballowe, Hannah Burnett, Kevin Clark, Anja Claus, Kate Cummings, Jon Daniels, Alana DeJoseph, Brooke Hecht, Bruce Jennings, Curt Meine, and Jeremy Ohmes. People of big ideas and bigger spirits, you make work a joy.

To the Wildings—Sara Beck, Sara Crosby, Margo Farnsworth, David Taylor, Missy Wick, and the not-forgotten A. C. Shilton. People of craft and care, you make my writing better.

To friends who encouraged the emergence of this book—John Hausdoerffer, Liam Heneghan, Christie Henry, Julian Hoffman, Lauren Markham, Lilian Pearce, Arthur Pearson, Bob Pyle, and Andreas Weber. People of skill and creativity, you nurture my thoughts.

To Keara McGraw—whose alchemical artistry brings Chicago's wildlife to life in these pages.

To the people and places that provided hospitality, a writing nest, and hummingbirds—Tony Anella and Cara McCulloch, who helped me feel at home during my writing residency at Aldo and Estella Leopold's cabin, "Mi Casita," in Tres Piedras, New Mexico. To Susan Tillett and the fine folks of Mesa Refuge, who provided a writing habitat on the edge of Tomales Bay in Point Reyes, California.

Portions of "A Companionable Dissolution to Plan A" and "The TV Graveyard near Tong's Tiki Hut" first appeared in *Orion* under the titles "Chicago, IL" and "City Creatures." "The Cool Red Eye of Chicago" first appeared in *Zoomorphic*. A number of these essays were fledged on the *City Creatures Blog*, an online publication of the Center for Humans and Nature. My thanks to these publications for providing cozy nests.

To the University of Chicago Press and the fine work of those who brought this book into its final form—Katherine Faydash, Miranda Martin, Alan Thomas, Ryan Li, Erin DeWitt, Melinda Kennedy, and Skye Agnew.

Coyote insisted I thank him. So, *thanks, Coyote*—and all the other animals who make Chicago sing.

NOTES

Prologue: A Companionable Dissolution to Plan A

1. Robert Michael Pyle, "A Nat'ral Histerrical Fella in an Unwondering Age," in *The Way of Natural History*, ed. Tom Fleischner (San Antonio, TX: Trinity University Press, 2011), 167.

2. Carl Smith, *The Plan of Chicago: Daniel Burnham and the Remaking of the American City* (Chicago: University of Chicago Press, 2006), xvi.

3. For a summary of both the extremity of change and the still-thriving biodiversity of the Chicagoland region, see Chicago Wilderness's *Atlas of Biodiversity*, https://archive.epa.gov/ecopage/web/pdf/chicago-wilderness-atlas-biodiversity-1997-65pp.pdf. For firsthand historical accounts of the landscape prior to extensive European settlement, as well as contemporary perspectives, see Joel Greenberg's *A Natural History of the Chicago Region* (Chicago: University of Chicago Press, 2002). For extensively researched insights regarding Chicago's role in initiating widespread ecological changes across the United States, see William Cronon's *Nature's Metropolis: Chicago and the Great West* (New York: Norton, 1991).

4. A deeper dive into Chicago's various phases of development, social and political unrest, changing demographics, and neighborhood-level revitalization and distress can be found in Larry Bennett, *The Third City: Chicago and American Urbanism* (Chicago: University of Chicago Press, 2010).

5. Perhaps one of Leopold's most admirable qualities was his willingness to challenge himself as a scientist and social commentator. In recent years, this

has become more apparent as a vast unpublished corpus of Leopold's writings became publicly available—first in anthologies, and now in a comprehensive online archive maintained by scholars affiliated with the University of Wisconsin. Valuable collections of Leopold's writings (some previously unpublished or very difficult to find) include Susan L. Flader and J. Baird Callicott, eds., *The River of the Mother of God and Other Essays by Aldo Leopold* (Madison: University of Wisconsin Press, 1991); David E. Brown and Neil Carmony, eds., *Aldo Leopold's Southwest* (Albuquerque: University of New Mexico Press, 1995); and J. Baird Callicott and Eric T. Freyfogle, eds., *For the Health of the Land: Previously Unpublished Essays and Other Writings* (Washington, DC: Island Press, 1999). The digitized archive of all Leopold writings, including his journals and an extensive body of correspondence, can be found at the website of the Aldo Leopold Archives, at http://digicoll.library.wisc.edu/AldoLeopold/.

6. Aldo Leopold, "Wilderness as a Form of Land Use," in *The River of the Mother of God and Other Essays by Aldo Leopold*, ed. Susan L. Flader and J. Baird Callicott (Madison: University of Wisconsin Press, 1991), 135; Aldo Leopold, *A Sand County Almanac and Sketches Here and There* (1949; New York: Oxford Press, 1989), 174.

7. Quotes from Aldo Leopold, *Sand County Almanac*, 224, viii–ix.

8. Dan Flores, *Coyote America: A Natural and Supernatural History* (New York: Basic Books, 2016), 25. Flores's accessible and engaging "biography" of coyotes (and Coyote) traces the intelligent and adaptable canids' evolutionary origins on the continent up to their present-day geographical explosion into every state (except Hawaii) and city. Those interested in coyote biology and sociality, as well as the lore associated with them, will find much in Flores's treatment to appreciate. For a collection of Coyote's exploits from various Native peoples, see the collection edited by Barry Lopez, *Giving Birth to Thunder, Sleeping with His Daughter: Coyote Builds North America* (1977; repr., New York: Harper Perennial, 2001). For tales that hew closer to a regional vision of Coyote, in this case interior Washington, see Mourning Dove's delightful recollections in Mourning Dove, *Coyote Stories*, ed. Heister Dean Guie, with notes by L. V. McWhorter (Old Wolf) and Jay Miller (Lincoln: University of Nebraska Press, 1990). For an erudite exploration of the trickster figure among various cultures, see Lewis Hyde, *Trickster Makes This World: Mischief, Myth, and Art* (New York: Farrar, Straus & Giroux, 1998).

9. I believe this first occurred to me when I read J. Donald Hughes's *Pan's Travail: Environmental Problems of the Ancient Greeks and Romans* (Baltimore: Johns Hopkins University Press, 1994). Hughes notes the deforestation described by Plato in the fourth century BCE, which led to ancient ports becoming landlocked by erosional sediments after hillsides swallowed once-bustling urban sites. For further examples of environmental trajectories gone wrong in relation to various civilizations, Jared Diamond's *Collapse: How Societies Choose to Fail or Succeed* (New York: Viking Penguin, 2005) should be all the fair warning anyone may require.

The Channel Coyotes

1. More on Stan and his work, including published scientific papers and the kinds of research he and his team do, can be found at the website of the Urban Coyote Research Project, at https://urbancoyoteresearch.com.

2. Donna Haraway, "Situated Knowledges: The Science Question in Feminism and the Privilege of Partial Perspective," *Feminist Studies* 14, no. 3 (Autumn 1988): 593–94.

3. Richard Nelson, *Make Prayers to the Raven: A Koyukon View of the Northern Forest* (Chicago: University of Chicago Press, 1983), 14.

4. Coyote stories were traditionally told in the wintertime, as Barre Toelken remarks, "*only* after the first killing frost and before the first thunderstorm— that is, in wintertime as defined by nature itself" (in Lopez, *Giving Birth to Thunder*, xiii). There are exceptions, according to Jay Miller, who notes that "the best Coyote stories are usually heard at wakes while family and friends are sitting up all night with the deceased. During the darkest of the night, old ladies will begin to tell the most outrageous stories, helping to relieve the grief and keep everyone awake" (in Mourning Dove, *Coyote Stories*, ix).

5. In Mourning Dove's telling of "The Spirit Chief Names the Animal People" in *Coyote Stories*, we get one perspective on the origins of this life-restoring power. After giving Coyote the ability to change into any form he wishes, the Spirit Chief declares: "To your twin brother, *Why-ay'-looh*, and to others I have given *shoo'-mesh* [medicine]. It is strong power. With that power Fox can restore your life should you be killed. Your bones may be scattered but, if there is one hair of your body left, Fox can make you live again. Others of the people can do the same with their *shoo'-mesh*" (23). For an example of Coyote putting himself back together after taking himself apart, see Lopez, *Giving Birth to Thunder*, 95.

6. Nor was this confined to coyotes. Wolves, mountain lions, and bears suffered tremendous losses across the United States, as did non-target populations of birds and small mammals who fed on poisoned carcasses and baits. In *Coyote America*, Flores recounts the justifications for such "control" efforts in some detail, as does Michael Robinson in *Predatory Bureaucracy: The Extermination of Wolves and the Transformation of the West* (Boulder: University Press of Colorado, 2005). Bounties and strychnine were the killing methods most utilized by stockmen in the nineteenth century, but by the early twentieth century, the government was manufacturing its own supply of strychnine, and coyote killing had been bureaucratized. By the mid-1920s, according to Flores, "this scorched-earth policy against coyotes yielded some 35,000 dead coyote bodies a year" (102–3). Once the science of ecology was better established and the justification for killing large- and medium-sized carnivores was shown to rest on ideological premises, voices raised against such indiscriminate killing began to swell. Yet the slaughter continued. From 1945 to 1971, 3.6 million coyote carcasses were "collected" by federal agents (Flores,

Coyote America, 147). Even today, as Flores notes, the federal agency Wildlife Services kills around eighty thousand coyotes a year, mostly on western rural lands (157, 177).

Scrapers of Sky

1. Quotes in this paragraph are from J. A. Baker, *The Peregrine* (New York: New York Review Books, 1967), 13–14.

2. For a nuanced and complicated history about the social dimensions of the restoration ecology movement, see William Jordan III and George Lubick, *Making Nature Whole: A History of Ecological Restoration* (Washington, DC: Island Press, 2011).

3. Michael Rosenzweig, *Win-Win Ecology: How the Earth's Species Can Survive in the Midst of Human Enterprise* (New York: Oxford University Press, 2003), 7.

4. Rosenzweig, *Win-Win Ecology*, 152.

Under Construction

1. Libby Hill, *The Chicago River: A Natural and Unnatural History* (Chicago: Lake Claremont Press, 2000), 139; and the revised edition of the same title (Carbondale: Southern Illinois University Press, 2016).

The TV Graveyard near Tong's Tiki Hut

1. The cameras provide monstrous amounts of data, too much for any one person or even a sizable staff to sift through. To meet demand, Seth and his team have partnered with the Adler Planetarium to engage citizens in identifying animals captured in the photographs. I'm told it's an addictive activity. Available at the Chicago Wildlife Watch website, http://www.chicagowildlifewatch.org.

2. E. O. Wilson, *The Creation: An Appeal to Save Life on Earth* (New York: Norton, 2006), 55; Michael E. Soulé, James A Estes, Brian Miller, and Douglas L. Honnold, "Strongly Interacting Species: Conservation Policy, Management, and Ethics," *BioScience* 55, no. 2 (February 2005): 175; and Reed F. Noss, "Is There a Special Conservation Biology?" *Ecography* 22 (1999): 113–22. For an expanded analysis of the ways in which scientists express forms of naturalistic spirituality, see Bron Taylor, *Dark Green Religion: Nature Spirituality and the Planetary Future* (Berkeley: University of California Press, 2010).

3. Henry D. Thoreau, *Walden: A Fully Annotated Edition*, ed. Jeffrey S. Cramer, annotated ed. (New Haven, CT: Yale University Press, 2004), 90.

4. Ibid., 50. There may be a coyote-related reason—buried deep in my subconscious—for why Thoreau popped into my head on my trip with Seth. When I was tracking down sources, I found a reference to Thoreau in one of my favorite Ed Abbey essays, "Down the River with Henry Thoreau." Abbey, who deeply admires Thoreau yet playfully scoffs at his high-minded and sometimes

puritanical goals, offers a load of vivid descriptors to describe him, including "the arrogant, insolent village crank," "a crusty character," "an unpeeled man," "a man with bark on him," and a compliment to Thoreau's thumbing his nose at conventional New England townie lifestyles: "Our suburban coyote."

De los pajaritos del monte

1. Readers may be pleased to know that in the interim between when I wrote this essay and when it was published, General Growth reduced itself to an acronym: GGP (General Growth Properties). At least there's been a de-growth in the amount of letters used in the company's name.

2. The examples in these paragraphs are from Ruben Cobos, *A Dictionary of New Mexico & Southern Colorado Spanish*, rev. and expanded ed. (Santa Fe: Museum of New Mexico Press, 2003). Cobos provides a nice overview of the "regional type" of Spanish spoken in rural New Mexico and southern Colorado, an "offshoot of the Spanish of northern Mexico" that "has survived by word of mouth for over four hundred years" (ix) because of its relative isolation from other Spanish-speaking areas.

3. Aaron Abeyta, *Rise, Do Not Be Afraid* (Denver, CO: Ghost Road Press, 2007), 20–22.

4. I draw here from Flores, who calls coyotes an "American original" who would make "a damned fine national totem" because of their unique history on the continent (*Coyote America*, 19). Flores also provides a glimpse into coyotes' "initial experiments with urban living" among the Aztecs (9–10), and he offers a helpful historical summary of early European and American explorers' observations and attempts to classify coyotes (see, in particular, 70–73).

An Etiquette of Sound

1. Jon Young, *What the Robin Knows: How Birds Reveal the Secrets of the Natural World* (New York: Houghton Mifflin Harcourt, 2012), 170, 69.

2. Young, *What the Robin Knows*, 31–32.

3. Young, *What the Robin Knows*, 62.

A Language That Transcends Words

1. Pablo Neruda, *Extravagaria*, trans. Alastair Reid (1958; New York: Farrar, Straus & Giroux, 1974), 63.

2. I first stumbled on the captivating story several years ago in *Utne Reader* magazine. For a fuller elaboration of the account, published posthumously, see Val Plumwood, *The Eye of the Crocodile* (Canberra, Australia: ANU Press, 2012).

3. Plumwood, *Eye of the Crocodile*, 14.

4. Plumwood, *Eye of the Crocodile*, 10.

5. Plumwood, *Eye of the Crocodile*, 16.

6. For further detail about the impact of this narrative, including the antecedents for the essay and its ongoing resonances in wolf and wildlife advocacy,

see Gavin Van Horn, "Fire on the Mountain: Ecology Gets Its Narrative Totem," *Journal for the Study of Religion, Nature and Culture* 5, no. 4 (2011): 437–64.

7. Quoted by Peter Atterton in his essay "Facing Animals," in *Facing Nature: Levinas and Environmental Thought*, ed. William Edelglass, James Hatley, and Christian Diehm (Pittsburgh, PA: Duquesne University Press, 2012), 28.

8. A letter that Leopold wrote to his mother, dated September 22, 1909, described the forest survey and noted the killing of two timber wolves. That the event did not leave much of an impression on the young Leopold—he spent more time lamenting the loss of his pipe, which he cheekily called the "greatest of sorrows"—reveals the degree of change in his later attitudes, which is of course well expressed in the final essay itself.

9. Leopold, *Sand County Almanac*, 129–30.

10. For more on this topic, see Diana L. Eck, *Darśan: Seeing the Divine Image in India*, 3rd ed. (New York: Columbia University Press, 1998).

11. Neil Shubin, *Your Inner Fish: A Journey into the 3.5-Billion-Year History of the Human Body* (New York: Vintage Books, 2008), 157.

The Cool Red Eye of Chicago

1. Ashley Pryor offers an astute and engaging tour of the relationship between Leopold's and Ouspensky's writing, especially their shared interest in an impersonal nature mysticism in "Thinking Like a Mystic: The Legacy of P. D. Ouspensky's *Tertium Organum* on the Development of Aldo Leopold's 'Thinking Like a Mountain,'" *Journal for the Study of Religion, Nature and Culture* 5, no. 4 (2011): 465–90.

2. Leopold, *Sand County Almanac*, 137–38.

3. Leopold, *Sand County Almanac*, 174.

Vulning

1. David Abram reflects on this tendency in "On Being Human in a More-Than-Human World," at the Center for Humans and Nature website, http://www.humansandnature.org/to-be-human-david-abram. See also his remarkable book *The Spell of the Sensuous* (New York: Pantheon Books, 1996). Abram coined the term *more-than-human world* as a way of designating the lateral intersubjective relationships that humans share with other earthen beings and forces, especially nonhuman animals. Employing a phenomenological approach (and a sometimes very personal narrative) to the study of religion, Abram argues that humans are "tuned for relationship" by the body's senses and dependent on the more-than-human world for our sense of identity: "The simple premise of this book is that we are human only in contact, and conviviality, with what is not human" (ix). Yes, indeed.

2. Nurit Bird-David, "'Animism' Revisited: Personhood, Environment, and Relational Epistemology," *Current Anthropology* 40 (1999): 67–91.

3. Graham Harvey's book *Animism: Respecting the Living World* (New York: Columbia University Press, 2005) provides an excellent overview of this topic,

past and present. For a more personal narrative, from the perspective of a scientist who also honors Potawatomi language and knowledge, including a practice of deep listening to nonhuman relations, see Robin Kimmerer's *Braiding Sweetgrass: Indigenous Wisdom, Scientific Knowledge, and the Teachings of Plants* (Minneapolis: Milkweed, 2013).

4. Quotations from Trombulak appear in his essay "Becoming a Neighbor," in *The Way of Natural History*, ed. Thomas Lowe Fleischner (San Antonio, TX: Trinity University Press, 2011), 131, 133.

The City Bleeds Out (Reflections on Lake Michigan)

1. Lao Tzu, *Tao Te Ching*, trans. Jonathan Star (New York: Jeremy P. Tarcher/Penguin, 2001), verse 78.

2. Lao Tzu, *Tao Te Ching*, verse 32.

3. Lao Tzu, *Tao Te Ching*, verse 4.

4. Ravens, for example, have captured a great deal of attention from cultures around the world. See Eric Mortensen, "Raven Augury from Tibet to Alaska: Dialects, Divine Agency, and the Bird's-Eye View," in *A Communion of Subjects: Animals, Science, and Ethics*, ed. Paul Waldau and Kimberley Patton (New York: Columbia University Press, 2006), 422–36. To delve deeper into avian esoterica, see Adele Nozedar, *The Secret Language of Birds: A Treasury of Myths, Folklore & Inspirational Stories* (London: HarperElement, 2006).

5. J. Freedman, "Iban Augury," *Bijdragen tot de Taal-, Land- en Volkenkunde* 117, no. 1 (1961): 141–67. According to Freedman, while other birds carry cultural significance, seven are most important to the Iban: rufous piculet, banded kingfisher, scarlet-rumped trogon, Diard's trogon, crested jay, maroon woodpecker, and white-rumped shama. Each has personality qualities unique to the bird and an elaborate symbology of associations related to color, habitual behavior, and types of vocalizations. "When any of these augural birds is seen or heard (so runs the theory of Iban augury) it can be assumed that the gods have something to communicate to man, for these birds, say the augurs, never reveal themselves without cause, they always have something to tell us" (147).

6. Freedman, "Iban Augury," 158.

7. Lao Tzu, *Tao Te Ching*, verse 8.

Great Blue Meditation

1. Quoted by Peter Matthiessen in *Nine-Headed Dragon River: Zen Journals, 1969–1982* (Boston: Shambhala, 1985), 14.

2. D. T. Suzuki, *Introduction to Zen Buddhism* (New York: Grove Press, 1964), 21.

A Question of Monarchs

1. See Lincoln P. Brower and Stephen B. Malcolm, "Animal Migrations: Endangered Phenomena," *American Zoologist* 31 (1991): 265–76; and Lincoln P.

Brower, Orley R. Taylor, Ernest H. Williams, Daniel A. Slayback, Raul R. Zubieta, and M. Isabel Ramírez, "Decline of Monarch Butterflies Overwintering in Mexico: Is the Migratory Phenomenon at Risk?" *Insect Conservation and Diversity* 5 (2012): 95–100

2. Lincoln P. Brower, "Canary in the Cornfield: The Monarch and the Bt Corn Controversy," *Orion Magazine* 20 (2001): 32–41; and Gary Paul Nabhan, "The Farmland Decline of Monarch Butterflies and Bees: What Do Our Food Choices Have to Do with It?," http://makewayformonarchs.org/i/archives/919.

3. In Chicago, there are signs of hope. Recognizing the opportunities for collaboration in urban areas, the Keller Science Action Center at the Field Museum partnered with the Landscape Conservation Cooperative Network and the US Fish and Wildlife Service to produce the Urban Monarch Conservation Guidebook (https://www.fieldmuseum.org/file/862866). The Field Museum offers additional resources about monarchs, including online story and coloring books for children in English and Spanish. Concerted efforts are under way to plant milkweed and establish monarch butterfly gardens throughout Chicago, as well as in other cities along the migratory flyway, such as St. Paul–Minneapolis, Kansas City, and Austin.

Shagbark Thoughts

1. Robert Michael Pyle, *The Thunder Tree: Lessons from an Urban Wildland* (New York: Houghton Mifflin, 1993), 148.

2. Pyle, *Thunder Tree*, 147.

3. Pyle, *Thunder Tree*, 147.

Vole-a-Thon

1. Allison has since taken a job with the Chicago Academy of Sciences at the Peggy Notebaert Nature Museum, where she is the curator of herpetology.

Desire Lines

1. Newhouse quoted in Robert H. Busch, *The Wolf Almanac* (Guilford, CT: Lyons Press, 1998), 121.

2. Henry David Thoreau, *Walden*, ed. Jeffrey S. Cramer, annotated ed. (New Haven, CT: Yale University Press, 2004), 291–92.

3. Leopold, *Sand County Almanac*, 96.

Greenways

1. Tim Ingold, *Being Alive: Essays on Movement, Knowledge and Description* (New York: Routledge, 2011), 44.

2. Whet Moser, "The 606 Shows How to Design a Park in the 21st Century (and Beyond)," *Chicago Magazine*, June 5, 2015, http://www.chicagomag.com/city-life/June-2015/The-606-Park-Design/.

3. Ingold, *Being Alive*, 148. Ingold's book is a masterwork of reflection on the phenomenon of movement and its relation to knowledge, especially in terms of its critique of Western societies' tendency to privilege "head over heels," operating under the assumption that elevation, detachment, and objectivity are the most legitimate ways of knowing the world. Good anthropologist that he is, in contrast Ingold highlights more grounded ways of knowing that bind together people, place, and story.

4. Although there are many sources to find statistics on automobiles' devastating costs to life, limb, and climate, Jeff Speck's *Walkable City: How Downtown Can Save America, One Step at a Time* (New York: North Point Press, 2011) details the wide societal impacts routinely accepted as collateral damage in an automotive culture. See in particular the chapter "Why Johnny Can't Walk." I diverge with some of his views about urban nature, but Speck makes a compelling and practical case for privileging urban walkability.

5. Ingold explores how this mode of locomotion is characteristic of an animic ontology, which is not about occupation (the claiming, dominance, and presumed ownership of "empty" places so characteristic of modernity) but about inhabitation of the world: "This tangle is the texture of the world. In the animic ontology, beings do not simply occupy the world, they *inhabit* it, and in so doing—in threading their own paths through the meshwork they contribute to its ever-evolving weave" (*Being Alive*, 71). Our lives are fluidly connected to those of others, and the comminglings (a meshwork of knotty entanglements, as Ingold might put it) of our paths with other agentive beings thicken the relationships that create a sense of place. For a remarkable essay that addresses how language can reflect this sense of inhabitation, see "Learning the Grammar of Animacy," in Robin Kimmerer's *Braiding Sweetgrass: Indigenous Wisdom, Scientific Knowledge, and the Teachings of Plants* (Minneapolis: Milkweed, 2013).

6. For engaging examples of this emerging literature, see Rob Cowen, *Common Ground: Encounters with Nature at the Edges of Life* (Chicago: University of Chicago Press, 2015); Nathanael Johnson, *Unseen City: The Majesty of Pigeons, the Discreet Charm of Snails & Other Wonders of the Urban Wilderness* (New York: Rodale, 2016); and Lyanda Lynn Haupt, *The Urban Bestiary: Encountering the Everyday Wild* (New York: Little, Brown, 2013).

7. For more on Burnham, and the aspirations of those involved in crafting this landmark document, see Smith, *Plan of Chicago.*

8. In his posthumously published book *The Prairie of Illinois Country* (Westmont, IL: DPM Ink, 2011), Robert Betz, a former biology professor at Northeastern Illinois University, attributes the growth of his "prairie fever" to stumbling upon "pioneer cemeteries" while on the search for prairie plants in the early 1960s. These small country plots, protected from the encroachment of surrounding cornfields, held historic botanical treasures. Inspired by the restoration possibilities and armed with little more than an atlas and a keen eye, Betz tracked down dozens of these remnant cemetery prairies. He then

arranged speaking engagements with members of cemetery administrative boards, highlighting the importance of the remnant "pioneer" prairies. He also gave talks at local high schools, advocating for the cemeteries' restoration and care. He happily notes that many of these sites eventually became Illinois and Indiana nature preserves.

9. See Natalie Moore, "New Redlining Maps Show Chicago Housing Discrimination," WBEZ, October 28, 2016, https://www.wbez.org/shows/wbez -news/new-redlining-maps-show-chicago-housing-discrimination/37c0dce7 -0562-474a-8e1c-50948219ecbb. For an excellent study of this devastating phenomenon, see Nathan McClintock, "From Industrial Garden to Food Desert: Demarcated Devaluation in the Flatlands of Oakland, California," in *Cultivating Food Justice: Race, Class, and Sustainability*, ed. Alison Hope Alkon and Julian Agyeman (Cambridge, MA: MIT Press, 2011), 89–120.

10. Quotes in this paragraph are from Rebecca Solnit, *Wanderlust: A History of Walking* (New York: Penguin Books, 2000), 11–12.

11. For the full essay, see Gary Snyder, *The Practice of the Wild* (Berkeley, CA: North Point Press, 1990), 78–96.

12. Robert Macfarlane, *The Old Ways: A Journey on Foot* (New York: Penguin Books, 2012), 27–31.

13. Solnit, *Wanderlust*, 29.

14. Snyder, *Practice of the Wild*, 94.

Blueways

1. Several portions of this chapter, including the reference to Jolliet here, are informed by Libby Hill's book, *The Chicago River: A Natural and Unnatural History* (Chicago: Lake Claremont Press, 2000), and the revised edition of the same title (Carbondale: Southern Illinois University Press, 2016).

2. The photograph came to my attention through Michael Bryson, who tracks the historical abuse of the creek and its value as present-day urban wilderness in "Canoeing through History: Wild Encounters on Bubbly Creek," in *City Creatures: Animal Encounters in the Chicago Wilderness*, ed. Gavin Van Horn and Dave Aftandilian (Chicago: University of Chicago Press, 2015).

3. Hill, *Chicago River*, vii.

4. David Ulin, *Sidewalking: Coming to Terms with Los Angeles* (Oakland: University of California Press, 2015), 23.

5. Theodore Schwenk, *Sensitive Chaos: The Creation of Flowing Forms in Water and Air*, trans. Olive Whicher and Johanna Wrigley (Letchworth, UK: Garden City Press, 1965), 14–15.

6. Schwenk, *Sensitive Chaos*, 18.

7. Schwenk, *Sensitive Chaos*, 84.

Mindways

1. William Cronon, *Nature's Metropolis: Chicago and the Great West* (New York: Norton, 1991), 18.

2. Smith, *Plan of Chicago*, 6.

3. Smith, *Plan of Chicago*, 37.

4. Leopold, *Sand County Almanac*, 224–25.

5. Leopold, *Sand County Almanac*, 204.

6. Paul Shepard, *The Others: How Animals Made Us Human* (Washington, DC: Island Press, 1996), 4.

7. For an essential source on the dustup, see Ralph H. Lutts, *The Nature Fakers: Wildlife, Science and Sentiment* (Charlottesville: University Press of Virginia, 1990). For a representative collection of the original stories, with commentary from Lutts, see Ralph H. Lutts, ed., *The Wild Animal Story* (Philadelphia: Temple University Press, 1998).

8. Leopold, *Sand County Almanac*, 204.

9. Aldo Leopold, "The Ecological Conscience," 1947, in *The River of the Mother of God and Other Essays by Aldo Leopold*, ed. Susan L. Flader and J. Baird Callicott (Madison: University of Wisconsin Press, 1991), 346.

Epilogue: Postscript to a Hope

The title of this epilogue comes from a line in Leopold's less-than-one-page essay "Draba," which appears in *A Sand County Almanac*. In the short reflection, he writes:

> Within a few weeks now Draba, the smallest flower that blows, will sprinkle every sandy place with small blooms. He who hopes for spring with upturned eye never sees so small a thing as Draba. He who despairs of spring with downcast eye steps on it, unknowing. He who searches for spring with his knees in the mud finds it, in abundance.... Botany books give it two or three lines, but never a plate or portrait. Sand too poor and sun too weak for bigger, better blooms are good enough for Draba. After all it is no spring flower, but only a postscript to a hope. (1987 [1949]: 26)

I've always been intrigued by that turn of phrase, "postscript to a hope," as though the humble flower were whispering, *Psst, spring has come again. Thought you should know.* Such a "postscript" struck me as a fitting title for an epilogue, a last word that signals the very beginning, old knowledge rising again to carry us forward into a new cycle.

1. Leopold, *Sand County Almanac*, 178.

2. Aldo Leopold, *Game Management* (repr.; Madison: University of Wisconsin Press, 1986), 423.

3. The reference here to frying a small fish comes from verse 60 in the *Tao Te Ching*, where this action is likened to the best way to govern. When I first read this line, I was completely baffled. So baffled that I couldn't get it out of my head. As a metaphor, and in keeping with Taoist philosophy, I think the admonition means that one should handle governance carefully, allowing people to pursue their own ends without too much interference. Don't fiddle

too much with the fish, just let the pan and the butter do the work. I thought it would be a hoot to have Lao Tzu literally frying a fish.

4. Lao Tzu, *Tao Te Ching*, verse 3.

5. Lao Tzu, *Tao Te Ching*, verse 2.

6. Lao Tzu, *Tao Te Ching*, verses 8–9.

7. Leopold, *Sand County Almanac*, 200.

8. Aldo Leopold, "Some Fundamentals of Conservation in the Southwest," in *The River of the Mother of God and Other Essays by Aldo Leopold*, ed. Susan L. Fladler and J. Baird Callicott (Madison: University of Wisconsin Press, 1991), 95.

9. Lao Tzu, *Tao Te Ching*, verse 31.

10. Lao Tzu, *Tao Te Ching*, verse 32.

11. Lao Tzu, *Tao Te Ching*, verse 21.

12. Lao Tzu, *Tao Te Ching*, verse 71.

13. Lao Tzu, *Tao Te Ching*, verse 63.

www.ingramcontent.com/pod-product-compliance
Lightning Source LLC
Chambersburg PA
CBHW032131020426
42334CB00016B/1121